MATERIALS SCIENCE AND ENGINEERING

Physiochemical Concepts,
Properties, and Treatments

Volume 2

MATERIALS SCIENCE AND ENGINEERING

Physiochemical Concepts,
Properties, and Treatments

Volume 2

Edited by
**Gennady E. Zaikov, DSc, A. K. Haghi, PhD,
and Ewa Kłodzińska, PhD**

Apple Academic Press

TORONTO   NEW JERSEY

Apple Academic Press Inc.	Apple Academic Press Inc.
3333 Mistwell Crescent	9 Spinnaker Way
Oakville, ON L6L 0A2	Waretown, NJ 08758
Canada	USA

©2015 by Apple Academic Press, Inc.

First issued in paperback 2021

Exclusive worldwide distribution by CRC Press, a member of Taylor & Francis Group
No claim to original U.S. Government works

ISBN 13: 978-1-77463-091-4 (pbk)
ISBN 13: 978-1-77188-009-1 (hbk)

Library of Congress Control Number: 2014937943

Library and Archives Canada Cataloguing in Publication

Materials science and engineering.

Includes bibliographical references and index.
Contents: Volume 1. Physical process, methods, and models/edited by Abbas Hamrang, PhD; Gennady E. Zaikov, DSc, and A.K. Haghi, PhD, Reviewers and Advisory Board Members -- Volume 2. Physiochemical concepts, properties, and treatments/edited by G.E. Zaikov, DSc, A.K. Haghi, PhD, and E. Kłodzińska PhD.
ISBN 978-1-77188-054-1 (set).--ISBN 978-1-77188-000-8 (v. 1 : bound).--
ISBN 978-1-77188-009-1 (v. 2 : bound)

1. Materials. I. Hamrang, Abbas, editor II. Zaikov, G. E. (Gennady Efremovich), 1935-, author, editor III. Haghi, A. K., author, editor IV. Kłodzińska, Ewa, editor V. Title: Physical process, methods, and models. VI. Title: Physiochemical concepts, properties, and treatments.

| TA403.M38 2014 | 620.1'1 | C2014-902356-1 |

Apple Academic Press also publishes its books in a variety of electronic formats. Some content that appears in print may not be available in electronic format. For information about Apple Academic Press products, visit our website at **www.appleacademicpress.com** and the CRC Press website at **www.crcpress.com**

ABOUT THE EDITORS

Gennady E. Zaikov, DSc

Gennady E. Zaikov, DSc, is Head of the Polymer Division at the N. M. Emanuel Institute of Biochemical Physics, Russian Academy of Sciences, Moscow, Russia, and Professor at Moscow State Academy of Fine Chemical Technology, Russia, as well as Professor at Kazan National Research Technological University, Kazan, Russia. He is also a prolific author, researcher, and lecturer. He has received several awards for his work, including the Russian Federation Scholarship for Outstanding Scientists. He has been a member of many professional organizations and is on the editorial boards of many international science journals.

A. K. Haghi, PhD

A. K. Haghi, PhD, holds a BSc in urban and environmental engineering from University of North Carolina (USA); a MSc in mechanical engineering from North Carolina A&T State University (USA); a DEA in applied mechanics, acoustics and materials from Université de Technologie de Compiègne (France); and a PhD in engineering sciences from Université de Franche-Comté (France). He is the author and editor of 65 books as well as 1000 published papers in various journals and conference proceedings. Dr. Haghi has received several grants, consulted for a number of major corporations, and is a frequent speaker to national and international audiences. Since 1983, he served as a professor at several universities. He is currently Editor-in-Chief of the *International Journal of Chemoinformatics and Chemical Engineering* and *Polymers Research Journal* and on the editorial boards of many international journals. He is also a member of the Canadian Research and Development Center of Sciences and Cultures (CRDCSC), Montreal, Quebec, Canada.

Ewa Kłodzińska, PhD

Ewa Kłodzińska, PhD, holds a PhD from Nicolaus Copernicus University, Faculty of Chemistry in Torun, Poland. For ten years, she has been doing research on determination and identification of microorganisms using the electromigration techniques for the purposes of medical diagnosis. Currently she is working at the Institute for Engineering of Polymer Materials and Dyes and investigates surface characteristics of biodegradable polymer material on the basis of zeta potential measurements. She has written several original articles, monographs, and chapters in books for graduate students and scientists. She has made valuable contributions to the theory and practice of electromigration techniques, chromatography, sample preparation, and application of separation science in pharmaceutical and medical analysis. Dr. Ewa Kłodzińska is a member of editorial boards of *ISRN Analytical Chemistry* and the *International Journal of Chemoinformatics and Chemical Engineering* (IJCCE).

CONTENTS

LIST OF CONTRIBUTORS

D. S. Andreev
Volgograd State Architect-build University, Sebrykov Department Michurin Street 21, Michailovka, Volgograd region, Russia

Yu. S. Artemova
Volgograd State Architect-build University, Sebrykov Department Michurin Street 21, Michailovka, Volgograd region, Russia

V. A. Babki
Volgograd State Architect-Build University Sebrykov Department, 403300 Michurin Street 21, Michailovka, Volgograd region, Russia, E-mail: sfi@reg.avtlg.ru

Heinrich Badenhorst
SARChI Chair in Carbon Materials and Technology, Department of Chemical Engineering, University of Pretoria, Lynwood Road, Pretoria, Gauteng, 0002, South Africa, P.O. Box 66464, Highveld Ext. 7, Centurion, Gauteng, 0169, South Africa, Tel.: +27 12 420 4173; Fax: +27 12 420 2516; E-mail: heinrich.badenhorst@up.ac.za

L. I. Bazylyak
Physical Chemistry of Combustible Minerals DepartmentInstitute of Physical–Organic Chemistry & Coal Chemistry named after L. M. LytvynenkoNational Academy of Sciences of Ukraine3^a Naukova Str., Lviv–53, 79053, UKRAINEe–mail: andriy_kytsya@yahoo.com

V. U. Dmitriev
Volgograd State Architect-Build University Sebrykov Departament, 403300 Michurin Street 21, Michailovka, Volgograd region, Russia, E-mail: sfi@reg.avtlg.ru

I. V. Dolbin
Kabardino-Balkarian State University, Nal'chik – 360004, Chernyshevsky st., 173, Russian Federation, E-mail: I_dolbin@mail.ru

K. Z. Gumargalieva
N.N.Semenov Institute of Chemical Physics, RAS, 4 Kosygin Street

H. Hlídková
Institute of Macromolecular Chemistry Academy of Sciences of the Czech Republic, v.v.i. Heyrovský Sq. 2, 162 06 Prague 6, Czech Republic

D. Horák
Institute of Macromolecular Chemistry Academy of Sciences of the Czech Republic, v.v.i. Heyrovský Sq. 2, 162 06 Prague 6, Czech Republic, E-mail: horak@imc.cas.cz

Yu. M. Hrynda
Physical Chemistry of Combustible Minerals Department, Institute of Physical Organic Chemistry and Coal Chemistry, NAS of Ukraine, 3a Naukova Str.Lviv, 79053, E-mail: Ukraineandriy_kytsya@yahoo.com

A. N. Inozemtsev
M. V. Lomonosov MSU, Biological Faculty, Leninskie Gory, 119991 Moscow, Russia, E-mail: olga-karp@newmail.ru

Yu. Kalashnikova
Volgograd State Architect-build University, Sebrykov Department Michurin Street 21, Michailovka, Volgograd region, Russia

O. V. Karpukhina
N. N. Semenov Institute of Chemical Physics, RAS, 4 Kosygin Street

S. V. Kolesov
The Institute of Organic Chemistry of the Ufa Scientific Centre the Russian Academy of Science, October Prospect 71, 450054 Ufa, Russia

G. V. Kozlov
Kabardino-Balkarian State University, Nal'chik – 360004, Chernyshevsky st., 173, Russian Federation

E. I. Kulish
Bashkir State UniversityRussia, Republic of Bashkortostan, Ufa, 450074, ul. Zaki Validi, 32Tel.: +7 (347) 229 96 14, E-mail: alenakulish@rambler.ru

A. R. Kytsya
Physical Chemistry of Combustible Minerals Department, Institute of Physical Organic Chemistry and Coal Chemistry, NAS of Ukraine, 3a Naukova Str.Lviv, 79053, E-mail: Ukraineandriy_kytsya@yahoo.com

Hieronim Maciejewski
Poznań Science and Technology Park of Adam Mickiewicz University Foundation, Rubież 46, 61-612 Poznań, Poland

Przemysław Pietras
Faculty of Chemistry, Adam Mickiewicz University, Umultowska 89b, 61-614 Poznań, Poland Zenon Foltynowicz
Faculty of Commodity Science, Poznan University of Economics, Al. Niepodległości 10, 61-875 Poznań, Poland

V. V. Podmasteryev
N. M. Emanuel Institute of Biochemical Physics, Russian Academy of Sciences, 4 Kosigin str. Moscow 119334, Russia

S. D. Razumovsky
N. M. Emanuel Institute of Biochemical Physics, Russian Academy of Sciences, 4 Kosigin str. Moscow 119334, Russia

O. V. Reshetnyak
Department of chemistry, Army Academy named after Hetman Petro Sahaydachnyi, 32 Hvardiyska Str., Lviv, 79012, Ukraine, Physical Chemistry of Combustible Minerals Department, Institute of Physical Organic Chemistry and Coal Chemistry, NAS of Ukraine, 3a Naukova Str.Lviv, 79053, Ukraineandriy_kytsya@yahoo.com

Jozef Richert
Institut Inzynierii Materialow Polimerowych I Barwnikow, 55 M. Sklodowskiej-Curie str., 87-100 Torun, Poland, E-mail: j.richert@impib.pl

G. A. Savin
Volgograd State Pedagogical University, 40013, Lenin Street, 27, Volgograd, Russia, E-mail: gasavin@mail.ru

A. S. Shurshina
Bashkir State University, Russia, Republic of Bashkortostan, Ufa, 450074, ul. Zaki Validi, 32; Tel.: +7 (347) 229 96 14, E-mail: alenakulish@rambler.ru

D. V. Sivovolov
Volgograd State Architect-build University, Sebrykov Departament Volgograd State Architect-build University, Sebrykov Departament

O. V. Stoyanov
Kazan National Research Technological University, Kazan, Tatarstan, Russia, E-mail: OV_Stoyanov@mail.ru

E. S. Titova
Volgograd State Technical University, 40013, Lenin Street, 28, Volgograd, Russia

R. R. Usmanova
Ufa State technical university of aviation, Ufa, Bashkortostan, Russia, E-mail: Usmanovarr@mail.ru;

G. E. Zaikov
Institute of Biochemical Physics Russian Academy of Sciences; 117977 Kosigin Street 4, Moscow, Russia, E-mail: chembio@sky.chph.ras.ru

LIST OF ABBREVIATIONS

Ag–NPs	silver nanoparticles
AKDs	alkyl ketene dimmers
AM	antibiotic amikacin
AMS	salts – sulfate
ASA	active surface area
ASAs	alkenyl succinic anhydrides
ChT	chitosan
ChTA	chitosan acetate
EDMA	ethylene dimethacrylate
GPTSM	glycidoxypropyltrimethoxysilane
HEMA	hydroxyethyl methacrylate
HVSEM	high-vacuum scanning electron microscopy
LVSEM	low-vacuum scanning electron microscopy
MS	medicinal substance
NM	non-modified
NNG	natural source
NSG	synthetically produced material
Par	polyarylate
PB	International Paper POL, Poland
PC	polycarbonate
PHEMA	superporous poly(2-hydroxyethyl methacrylate
PIR	piracetam drug form
PMMA	poly(methyl methacrylate)
PP	Hoomark, Ltd., Jędrzychowice, Poland
PPD	phenylenediamines
RFL	third graphite
SEM	scanning electron microscopy
SPR	surface plasmon resonances
TEOS	tetraethoxysilane
TGA	thermogravimetric analyzer

TM	Blue Dolphin Tapes, Poland
TZ	Kreska, Bydgoszcz, Poland
U740	ethyl silicate
VOF	volume of fluid
CSC	stretched chains

LIST OF SYMBOLS

Q_{ir}	charge of a liquid on irrigating
ρ_l	fluid density
σ_f^c	fracture stress of composite
d_f^{cl}	nanocluster structure
σ_f^n	nominal (engineering) fracture stress
σ_f^m	polymer matrix
$\tilde{\varepsilon}_0$	proper strain of the loosely-packed matrix.
d_f^{cl}	real solids maximum dimension
Q_{cir}	the charge recirculation a liquid;
$[SO_2]_{gas}$	rate of release of volatile products
$[X]_\tau$	current concentration of isocyanate
b	Burgers vector
b - a	channel width
c	empirical coefficient of order
C_∞	characteristic ratio
d	dimension of Euclidean space
D_p	nanofiller particles diameter in nm
d_{surf}	nanocluster surface
d_u	accessible for contact
$E_{l.m}$	loosely-packed matrix elasticity modulus
E_n	elasticity moduli of nanocomposites
G	shear modulus
G_c	shear moduli of composite
$G_{l.m}$	shear moduli of polymer
G_p	strain hardening modulus
h	depth
h_g	fluid level
h_K	channel altitude
h_l	fluid level
K_s	stress concentration coefficient

K_T	bulk modulus
L	filler particle size
l	length of a crack
l_0	main chain skeletal length
l_k	specific spatial scale of structural changes
l_{st}	statistical segment length
M	repeated link molar mass,
m_∞	relative amount of water
m_0	initial mass of ChT in a film
M_{cl}	molecular weight of the chain part between cluster
M_e	molecular weight of chain
n	the number of cracks
N_A	Avogadro number
p	solid-state component
p_c	percolation threshold
pKa	universal index of acidity
q	parameter
$q \, \rho^{max}$	the greatest and the smallest density values
r	the radius of a particle
R	universal gas constant
R_{cl}	distance between nanoclusters
S	macromolecule cross
S_g	cross-section of the contact channel
S_r	cross-section of the contact channel
T	testing temperature
T_g	testing, glass transition
T_m	melting temperatures
W_c	optimum speed of gases
W_{max}	rate of the nucleus growth
W_r	relative speed of gases in the channel
α	the solubility coefficient
α_m	parameter
β_p	critical exponents
Δm	weight the absorbed film of water
ε_1	real part of the value of dielectric transmissivity
ε_2	imaginary part of the value of dielectric transmissivity

ε_f	strain at fracture
ε_m	dielectric constant of the surrounding medium
ε_M	dielectric transmissivity of the solvent
ε_Y	yield strain
η	exponent
q_{IP}	angle between the normal to IP
λ	length of a wave of the electromagnetic irradiation
λ_b	the smallest length of acoustic irradiation sequence
λ_k	length of irradiation sequence
ν	Poisson's ratio
ν_p	correlation length
ρ	polymer density
ρ	polymer density
ρ_{cl}	nanocluster density
ρ_d	density if linear
τ_{0Y}	theoretical value of the shear stress at yielding
τ_{in}	initial internal stress
φ_f	polymer matrix
φ_{if}	interfacial regions relative fraction
φ_n	nanofiller volume contents
χ	relative fraction of elastically
W_n	nanofiller mass contents

PREFACE

This volume has an important role in materials science and engineering on the macro and nanoscale. The book provides original, theoretical, and important experimental results. Some research uses non-routine methodologies often unfamiliar to some readers. Furthermore papers on novel applications of more familiar experimental techniques and analyses of composite problems are included.

This book brings together research contributions from eminent experts on subjects that have gained prominence in material and chemical engineering and science. It presents the last developments along with case studies, explanatory notes, and schematics for clarity and enhanced understanding.

Investigation on the influence of a strong electric field on the electrical, transport and diffusion properties of carbon nanostructures is discussed in chapter 1.

The purpose of chapter 2 is to study, by using DSC, the oxidation stability of PUE samples derived from butadiene and isoprene copolymer, and comparative assessment of OIT performance in the presence of different brands of pentaerythritol tetrakis[3-(3',5'-di-tert-butyl-4'-hydroxyphenyl) propionate] stabilizer.

The data on aromatic polyesters based on phthalic and n-oxybenzoic acid derivatives have been presented in chapter 3 along with various methods of synthesis of such polyesters developed by scientists from different countries for last 50 years

Polymerization of butadiene and isoprene under action of microheterogeneous titanium based catalyst with ultrasonic irradiation of the reaction mixture at the initial time is studied in chapter 4.

In chapter 5, a case study is presented about electric conductivity of polymer composites.

Chapter 6 presents the results of low-temperature, oxygen plasma activation of silica, kaolin and wollastonite. Fillers were modified in a tumbler reactor, enabling rotation of powders in order to modify their entire volume effectively.

Radiation crosslinking of elastomers has been receiving increasing attention. The reactions induced by high energy ionizing radiation are very complicated and the mechanisms still remain not entirely comprehended. Ionizing radiation crosslinking of acrylonitrile-butadiene rubber, filled with 40 phr of silica, with incorporated sulphur crosslinking system was the object of study. To investigate the influence of components such as sulphur and crosslinking accelerator-dibenzothiazole disulphide (DM) on the process, a set of rubber samples with various sulphur to crosslinking accelerator ratio was prepared and irradiated with 50, 122 and 198 kGy. Crosslink density and crosslink structure were analyzed and mechanical properties of the rubber samples were determined chapter 7.

In chapter 8, a case study investigates the sorption properties of biodegradable polymer materials.

Chapter 9 concerns questions of division of multicomponent solutions by means of polyamide membranes in the course of ultrafiltration. The question of influence of low-frequency fluctuations on a polyamide membrane for the purpose of increase of its productivity is also considered in this chapter.

Thermo-mechano-chemical changes of natural rubber SVR 3L under treatment internal mixer at selfheating have been studied in chapter 10. Effect of molecular mass and content of gel-fraction of natural rubber is shown as well. Properties of rubber compounds and vulcanized rubber are presented in this chapter.

Membrane filtration is an important technology for ensuring the purity, safety and/or efficiency of the treatment of water or effluents. In this study, various types of membranes are reviewed first. After that, the states of the computational methods are applied to membranes processes. Many studies have focused on the best ways of using a particular membrane process. But the design of new membrane systems requires a considerable amount of process development as well as robust methods. Monte Carlo and molecular dynamics methods can especially provide a lot of interesting information for the development of polymer/carbon nanotube membrane processes. A detailed review on polymer/carbon nanotube membrane filtration presented in the last chapter.

— **Gennady E. Zaikov, DSc, A. K. Haghi, PhD,**
and Ewa Kłodzińska, PhD

CHAPTER 1

THEORETICAL AND AN EXPERIMENTAL RESEARCH OF EFFICIENCY OF GAS PURIFICATION IN ROTOKLON WITH INTERNAL CIRCULATION OF A LIQUID

R. R. USMANOVA and G. E. ZAIKOV

CONTENTS

1.1   INTRODUCTION

For the machining of great volumes of an irrigating liquid and slurry salvaging the facility of bulky, capital-intensive, and difficult systems of water recycling are introduced. This process of clearing of gas considerably do a rise to its commensurable with clearing cost at application of the most difficult and cost intensive systems of dry clearing of gases (electrostatic precipitators and bag hoses) is required.

In this connection necessity for creation of such wet-type collectors which would work with the low charge of an irrigating liquid now has matured and combined the basic virtues of modern means of clearing of gases: simplicity and compactness, a high performance, a capability of control of processes of a dust separation and optimization of regimes.

To the greatest degree modern demands to the device and activity of apparatuses of clearing of industrial gases there match wet-type collectors with inner circulation the fluids gaining now more and more a wide circulation in systems of gas cleaning in Russia and abroad.

1.2   SURVEY OF KNOWN CONSTRUCTIONS OF SCRUBBERS WITH INNER CIRCULATION OF THE FLUID

An easy way to comply with the journal paper formatting requirements is to use this document as a template and simply type your text into it. The device and maintenance of systems of wet clearing of air are considerably facilitated, if water admission to contact zones implements as a result of its circulation in the apparatus. Slurry accumulating in it thus can continuously be retracted or periodically or by means of mechanical carriers, in this case necessity for water recycling system disappears, or a hydraulic path—a drain of a part of water. In the latter case, the device of system of water recycling can appear expedient, but load on it is much less, than at circulation of all volume of water [1, 2].

Dust traps of such aspect are characterized by presence of the capacity filled with water. Cleared air contacts to this water, and contact conditions are determined by interacting of currents of air and waters. The same interacting calls a water circulation through a zone of a contact at the expense of energy of the most cleared air.

The water discharge is determined by its losses on transpiration and with deleted slurry. At slurry removal by mechanical scraper carriers or manually the water discharge minimum also makes only 2–5 g on 1 $м^3$ air. At periodic drain of the condensed slurry the water discharge is determined by consistency of slurry and averages to 10 g on 1 $м^3$ air, and at fixed drain the charge does not exceed 100–200 g on 1 $м^3$ air. Filling of dust traps with water should be controlled automatically. Maintenance of a fixed level of water has primary value as its oscillations involve essential change as efficiency, and productivity of system.

The basic most known constructions of these apparatuses are introduced on Fig. 1.1 [3].

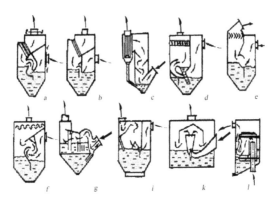

FIGURE 1.1   Constructions of scrubbers with inner circulation of a fluid: (a) rotoklon N (USA); (b) PVM CNII (Russia); (c) a scrubber a VNIIMT (Russia); (d) a dust trap to me (Czechoslovakia); (e) dust trap WNA (Germany); (f) dust trap "Asco" (Germany); (g) dust trap LGP (Russia); (i) dust trap "Klayrator" (USA); (k) dust trap VDN (Austria); (l) rotoklon RPA a NIIOGAS (Russia).

Mechanically each of such apparatuses consists of contact channel fractionally entrained in a fluid and the drip pan merged in one body. The principle of act of apparatuses is grounded on a way of intensive wash down of gases in contact channels of a various configuration with the subsequent separation of a water gas flow in the drip pan. The fluid which has thus reacted and separated from gas is not deleted at once from the apparatus, and circulates in it and is multiply used in dust removal process.

Circulation of a fluid in the wet-type collector is supplied at the expense of a kinetic energy of a gas flow. Each apparatus is supplied by the device for maintenance of a fixed level of a fluid, and also the device for removal of slurry from the scrubber-collecting hopper.

Distinctive features of apparatuses are:

1. Irrigating of gas by a fluid without use of injectors that allows using for irrigating a fluid with the high contents of suspended matters (to 250 mg/m³);

2. Landlocked circulation of a fluid in apparatuses which allows to reuse a fluid in contact devices of scrubbers and by that to device out its charge on clearing of gas to 0.5 kg/m³, that is, in 10 and more times in comparison with other types of wet-type collectors;

3. Removal of a collected dust from apparatuses in the form of dense with low humidity that allows to simplify dust salvaging to diminish load by water treating systems, and in certain cases in general to refuse their facility;

4. Layout of the drip pan in a body of the apparatus, which allows diminishing sizes of dust traps to supply their compactness.

The indicated features and advantages of such scrubbers have led to wide popularity of these apparatuses, active working out of various constructions, research and a heading of wet-type collectors, as in Russia, and abroad.

The scrubbers introduced on Fig. 1.1, concern to apparatuses with non-controllable operating conditions as in them there are no gears of regulating. In scrubbers of this type, the stable conditions of activity of a high performance are difficultly supplied, especially at varying parameters of cleared gas (pressure, temperature, a volume and dust content. In this connection wet scrubbers with controlled variables are safer and perspective. Regulating of operating conditions allows changing a hydraulic resistance from which magnitude, according to the power theory of a wet dust separation, efficiency of trapping of a dust depends. Regulating of parameters allows operating dust traps in an optimum regime at which optimum conditions of interacting of phases are supplied and peak efficiency of trapping of a dust with the least power expenditures is attained. Dust traps acquire the great value with adjustable resistance also for stabilization of processes of gas cleaning at varying parameters of cleared gas. A row of such scrubbers is introduced in Fig. 1.2.

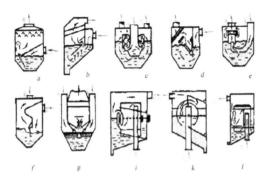

FIGURE 1.2 Apparatuses with controlled variables. (a) under the patent №1546651 (Germany), (b) the ACE №556824 (USSR), (c) the ACE № 598625 (USSR), (d) the ACE №573175 (USSR), (e) under the patent № 1903985 (Germany), (f) the ACE № 13686450 (France), (g) the ACE № 332845 (USSR), (i) the ACE № 318402 (USSR), (k) the ACE № 385598 (USSR), (l) type RPA a NIIOGAS (USSR).

The interesting principle of regulating is applied in the dust traps figured on Fig. 1.1, and 1.2. In these apparatuses, contact devices are had on a wall of the floating chamber entrained in a fluid and hardened in a body by means of joints. Such construction of dust traps allows supporting automatically to constants an apparatus hydraulic resistance at varying gas load.

From literary, data follows that known constructions of scrubbers with inner circulation of a fluid work in a narrow range of change of speed of gas in contact channels and are used in industrial production in the core for clearing of gases of a size dispersivity dust in systems of an aspiration of auxiliaries [3–5]. Known apparatuses are rather sensitive to change of gas load on the contact channel and to fluid level, negligible aberrations of these parameters from best values lead to a swing of levels of a fluid at contact channels, to unstable operational mode and dust clearing efficiency lowering. Because of low speeds of gas in contact channels known apparatuses have large gabarits. These deficiencies, and also a weak level of scrutiny of processes proceeding in apparatuses, absence of safe methods of their calculation hamper working out of new rational constructions of wet-type collectors of the given type and their wide heading in manufacture. In this connection necessity of more detailed theoretical and experimental study of scrubbers with inner circulation of a fluid for the purpose of the prompt use of the most effective and cost-effective constructions in systems of clearing of industrial gases has matured.

1.3   ARCHITECTURE OF HYDRODYNAMIC INTERACTING OF PHASES

In scrubbers with inner circulation of a fluid process of interacting of gas, liquid and hard phases in which result the hard phase (dust), finely divided in gas, passes in fluid implements. Because density of a hard phase in gas has rather low magnitudes (to 50 g/m³), it does not render essential agency on hydrodynamics of flows. Thus, hydrodynamics study in a scrubber with inner circulation of a fluid is reduced to consideration of interacting of gas and liquid phases.

Process of hydrodynamic interacting of phases it is possible to disjoint sequentially proceeding stages on the following:

- fluid acquisition by a gas flow on an entry in the contact device;
- fluid subdivision by a fast-track gas flow in the contact channel;
- integration of drops of a fluid on an exit from the contact device;
- branch of drops of a fluid from gas in the drip pan.

1.3.1   FLUID ACQUISITION BY A GAS FLOW ON AN ENTRY IN THE CONTACT DEVICE

Before an entry in the contact device of the apparatus there is a contraction of a gas flow to increase in its speed, acquisition of high layers of a fluid and its hobby in the contact channel. Functionability of all dust trap depends on efficiency of acquisition of a fluid a gas flow, without fluid acquisition will not be supplied effective interacting of phases in the contact channel and, hence, qualitative clearing of gas of a dust will not be attained. Thus, fluid acquisition by a gas flow on an entry in the contact device is one of defined stages of hydrodynamic process in a scrubber with inner circulation of a fluid. Fluid acquisition by a gas flow can be explained presence of interphase turbulence, which is advanced on an interface of gas and liquid phases. Conditions for origination of interphase turbulence is presence of a gradient of speeds of phases on boundaries, difference of viscosity of flows, and an interphase surface tension. At gas driving over a surface of a fluid the last will break gas boundary layers therefore in them there are the turbulent shearing stresses promoting cross-section transfer of energy. Originating cross-section turbulent oscillations lead to penetration of turbulent gas curls into boundary layers of a fluid with the subsequent illuviation of these stratums in curls. Mutual penetration of curls of boundary layers leads as though to the clutch of gas with a fluid on a phase boundary and to hobby of high layers of a fluid for moving gas over its surface. Intensity of such hobby depends on a kinetic energy of a gas flow, from its speed over a fluid at an entry in the contact device. At gradual increase in speed of gas there is a change of a surface of a fluid at first from smooth to undular, then ripples are organized and, at last, there is a fluid dispersion in gas. Mutual penetration of curls of boundary layers leads as though to the clutch of gas with a fluid on a phase boundary and to hobby of high layers of a fluid for moving gas over its surface. Intensity of such hobby depends on a kinetic energy of a gas flow, from its speed over a fluid at an entry in the contact device. The quantitative assessment of efficiency of acquisition in wet-type collectors with inner circulation of a fluid is expedient for conducting by means of a parameter $m=V_z/V_g$ m³/m³ equal to a ratio of volumes of liquid and gas phases in contact channels and characterizing the specific charge of a fluid on gas irrigating in channels. Obviously that magnitude m will be determined, first of all, by speed of a gas flow on an entry in the contact channel. Other diagnostic variable is fluid level on an entry in the contact channel, which can change cross-section of the channel and influence speed of gas:

$$\frac{v_r}{s_r} = \frac{V_r}{bh_k - bh_g} - \frac{v_r}{b(h_k - h_g)} \tag{1}$$

where, S_z is the cross-section of the contact channel; b is the channel width; h_K is the channel altitude; h_g is the fluid level.

Thus, for the exposition of acquisition of a fluid a gas flow in contact channels it is enough to gain experimental relation of following type:

$$m = f(W_r, h_{\text{ж}}) \tag{2}$$

1.3.2 FLUID SUBDIVISION BY A FAST-TRACK GAS FLOW IN THE CONTACT CHANNEL

As shown further, efficiency of trapping of corpuscles of a dust in many respects depends on a size of drops of a fluid: with decrease of a size of drops the dust clearing efficiency raises. Thus, the given stage of hydrodynamic interacting of phases is rather important.

Process of subdivision of a fluid by a gas flow in the contact channel of a dust trap occurs at the expense of high relative speeds between a fluid and a gas flow. For calculation of average diameter of the drops gained in contact channels, it is expedient to use the empirical formula of the Japanese engineers Nukiymas and Tanasavas, which allows considering agency of operating conditions along with physical performances of phases.

$$D_o = \frac{585 \cdot 10^3 \sqrt{\sigma}}{W_r} + 49,7 \left(\frac{\mu_l}{\sqrt{\rho_l \sigma_l}} \right)^{0,2} \frac{L_l}{V_r} \tag{3}$$

where, W_r is the relative speed of gases in the channel, m/s; σ_l is the factor of a surface tension of a fluid, N/m; ρ_l is the fluid density, kg/m³; μ_l is the viscosity of a fluid, the Pas/with; L_l is the volume-flow of a fluid, m³; V_r is the volume-flow of gas, m³.

In Fig. 1.3, computational curves of average diameter of drops of water in contact channels depending on speed of a gas flow are resulted. Calculation is conducted by Eq. (3) at following values of parameters: $\sigma = 720 \times 10^3$ N/m; $\rho_l = 1000$ kg/m³; $\mu = 1.01 \times 10^{-2}$ P/s.

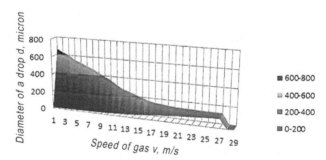

FIGURE 1.3 Relation of an average size of drops of water in blade impellers from speed of gas.

The gained relations testify that the major operating conditions on which the average size of drops in contact channels depends on the speed of gas flow W_r and the specific charge of a fluid on gas irrigating m. These parameters determine hydrodynamic structure of an organized water gas flow.

With growth of speed of gas process of subdivision of a fluid by a gas flow gains in strength, and drops of smaller diameter are organized. The most intensive agency on a size of drops renders change of speed of gas in the range from 7 to 20 m/s, at the further increase in speed of gas (>20 m/s) intensity of subdivision of drops is reduced. It is necessary to note that in the most widespread constructions of shock-inertial apparatuses (rotoklons N), which work at speed of gas in contact devices of 15 m/s, the size of drops in the channel is significant and makes 325–425 microns. At these operating conditions and sizes of drops qualitative clearing of gas of a mesh dispersivity dust is not attained. For decrease of a size of drops and raise of an overall performance of these apparatuses the increase in speed of gas to 30, 40, 50 m/s and more depending on type of a trapped dust is necessary.

The increase in the specific charge of a fluid at gas irrigating leads to the growth of diameter of organized drops. So, at increase m with 0.1×10^{-3} to 3×10 m³/m³ the average size of drops is increased approximately at 150 microns. For security of minimum diameter of drops in contact channels of shock-inertial apparatuses the specific charge of a fluid on gas irrigating should be optimized over the range $(0.1–1.5 \times 10$ m³/m³). It is necessary to note that in the given range of specific charges with a high performance the majority of fast-track wet-type collectors works.

1.3.3 INTEGRATION OF DROPS OF A FLUID ON AN EXIT FROM THE CONTACT DEVICE

On an exit from the contact device, there is an expansion of a water gas stream and integration of drops of a fluid at the expense of their concretion. The maximum size of the drops weighed in a gas flow, is determined by stability conditions: the size of drops will be that more than less speed of a gas flow. Thus, on an exit from the contact device together with fall of speed of a gas flow the increase in a size of drops will be observed. Turbulence in an extending part of a flow more than in the channel with fixed cross-section, and it grows with increase in an angle of jet divergence, and it means that speed of turbulent concretion will grow in an extending part of a flow also with increase in an angle of jet divergence. The more full there will be a concretion of corpuscles of a fluid, the drop on an exit from the contact device will be larger and the more effectively they will be trapped in the drip pan.

Practice shows that the size a coagulation of drops on an exit makes of the contact device, as a rule, more than 150 microns. Corpuscles of such size are easily trapped in the elementary devices (the inertia, gravitational, centrifugal, etc.).

1.3.4 BRANCH OF DROPS OF A FLUID FROM A GAS FLOW

The inertia and centrifugal drip pans are applied to branch of drops of a fluid from gas in shock-inertial apparatuses in the core. In the inertia drip pans the branch implements at the expense of veering of a water gas flow. Liquid drops, moving in a gas flow, possess definitely a kinetic energy thanks to which at veering of a gas stream they by inertia move rectilinearly and are inferred from a flow. If to accept that the drop is in the form of a sphere and speed of its driving is equal in a gas flow to speed of this flow the kinetic energy of a drop, moving in a flow, can be determined by:

$$E_\kappa = \frac{\pi D_0^{\ 3}}{6} \rho_l \frac{W^2_{\ r}}{2}$$

$$(4)$$

With the above formula, decrease of diameter of a drop and speed of a gas flow the drop kinetic energy is sharply diminished. At gas-flow deflection the inertial force forces to move a drop in a former direction. The more the drop kinetic energy, the is more and an inertial force.

$$E_\kappa = \frac{\pi D_0^{\ 3}}{6} \rho_l \frac{dW_r}{d\tau}$$

$$(5)$$

Thus, with flow velocity decrease in the inertia drip pan and diameter of a drop the drop kinetic energy is diminished, and efficiency drop spreads is reduced. However the increase in speed of a gas flow cannot be boundless as in a certain velocity band of gases there is a sharp lowering of efficiency drop spreads owing to origination of secondary ablation the fluids trapped drops. For calculation of a breakdown speed of gases in the inertia drip pans it is possible to use the formula, m/s:

$$W_c = K \sqrt{\frac{\rho_l - \rho_\kappa}{\rho_r}} \tag{6}$$

where, W_c is the optimum speed of gases in free cross-section of the drip pan, m/s; K is the the factor defined experimentally for each aspect of the drip pan.

Values of factor normally fluctuate over the range 0.1–0.3. Optimum speed makes from 3 to 5 m/s.

1.4 PURPOSE AND RESEARCH PROBLEMS

The following was the primal problems, which were put by working out of a new construction of the wet-type collector with inner circulation of a fluid:

- creation of a dust trap with a broad band of change of operating conditions and a wide area of application, including for clearing of gases of the basic industrial assemblies of a mesh dispersivity dust;
- creation of the apparatus with the operated hydrodynamics, allowing to optimize process of clearing of gases taking into account performances of trapped ingredients;
- to make the analysis of hydraulic losses in blade impellers and to state a comparative estimation of various constructions of contact channels of an impeller by efficiency of security by them of hydrodynamic interacting of phases;
- to determine relation of efficiency of trapping of corpuscles of a dust in a rotoklon from performance of a trapped dust and operating conditions major of which is speed of a gas flow in blade impellers. To develop a method of calculation of a dust clearing efficiency in scrubbers with inner circulation of a fluid.

1.5   EXPERIMENTAL RESEARCHES

1.5.1   THE EXPOSITION OF EXPERIMENTAL INSTALLATION AND THE TECHNIQUE OF REALIZATION OF EXPERIMENT

The rotoklon represents the basin with water on which surface on a connecting pipe of feeding into of dusty gas the dust-laden gas mix arrives. Over water surface gas deploys, and a dust contained in gas by inertia penetrate into a fluid. Turn of blades of an impeller is made manually, rather each other on a threaded connection by means of handwheels. The slope of blades was installed in the interval 25°– 45° to an axis.

In a rotoklon three pairs lobes sinusoidal a profile, the regulations of their rule executed with a capability are installed. Depending on cleanliness level of an airborne dust flow the lower lobes by means of handwheels are installed on an angle defined by operational mode of the device. The rotoklon is characterized by presence of three slotted channels, a formation the overhead and lower lobes, and in everyone the subsequent on a course of gas the channel the lower lobe is installed above the previous. Such arrangement promotes a gradual entry of a water gas flow in slotted channels and reduces thereby a device hydraulic resistance. The arrangement of an input part of lobes on an axis with a capability of their turn allows creating a diffusion reacting region. Sequentially had slotted channels create in a diffusion zone organized by a turn angle of lobes, a hydrodynamic zone of intensive wetting of corpuscles of a dust. In process of flow moving through the fluid-flow curtain, the capability of multiple stay of corpuscles of a dust in hydrodynamically reacting region is supplied that considerably raises a dust clearing efficiency and ensures functioning of the device in broad bands of cleanliness level of a gas flow.

The construction of a rotoklon with adjustable sinusoidal lobes is developed and protected by the patent of the Russian Federation, capable to solve a problem of effective separation of a dust from a gas flow [6]. Thus water admission to contact zones implements as a result of its circulation in the apparatus.

The rotoklon with the adjustable sinusoidal lobes, introduced in Fig. 1.4 contains a body (3) with connecting pipes for an entry (7) and an exit (5) gases in which steams of lobes sinusoidal a profile are installed. Moving of the overhead lobes (2) implements by means of screw jacks (6), the lower lobes (1) are fixed on an axis (8) with a capability of their turn. The turn angle of the lower lobes is chosen from a condition of a persistence of speeds of an airborne dust flow. For regulating of a turn angle output parts of the lower lobes (1) are envisioned handwheels. Quantity of pairs lobes is determined by productivity of the device and cleanliness level of an airborne dust flow that is a regime of a stable running of the device. In the lower part of a body there is a connecting pipe for a drain of slime water (9). Before a connecting pipe for a gas make (5) the labyrinth drip pan (4) is installed. The rotoklon works as follows. Depending on cleanliness level of an airborne dust flow the overhead lobes (5) by means of screw jacks (6), and the lower lobes (1) by means of handwheels are installed on an angle defined by operational mode of the device. Dusty gas arrives in the upstream end (7) in a top of a body (3) apparatuses. Hitting about a fluid surface, it changes the direction and passes in the slotted channel organized overhead (2) and lower (1) lobes. Thanks to the driving high speed, cleared gas captures a high layer of a fluid and atomizes it in the smallest drops and foam with an advanced surface. After consecutive transiting of all slotted channels gas passes through the labyrinth drip pan (4) and through the discharge connection (5) is deleted in an aerosphere. The collected dust settles out in the loading pocket of a rotoklon and through a connecting pipe for a drain of slime water (9), together with a fluid, is periodically inferred from the apparatus.

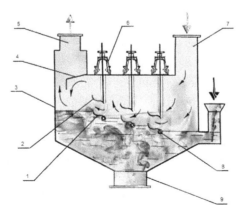

FIGURE 1.4 A rotoklon general view.

Lower (1) and the overhead (2) lobes; a body (3); the labyrinth drip pan (4); connecting pipes for an entry (7) and an exit (5) gases; screw jacks (6); an axis (8); a connecting pipe for a drain of slurry (9).

Noted structural features do not allow using correctly available solutions on hydrodynamics of dust-laden gas flows for a designed construction. In this connection, for the well-founded exposition of the processes occurring in the apparatus, there was a necessity of realization of experimental researches.

Experiments were conducted on the laboratory-scale plant "rotoklon" introduced in Fig. 1.5.

The examined rotoklon had three slotted channels speed of gas in which made to 15 km/s. At this speed the rotoklon had a hydraulic resistance 800 passes. Working in such regime, it supplied efficiency of trapping of a dust with input density 0.5 g/nm³ and density 1200 kg/m³ at level of 96.3% [7].

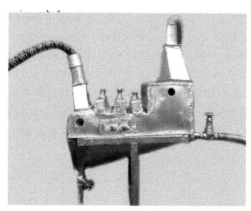

FIGURE 1.5   Experimental installation "rotoklon."

In the capacity of modeling system air and a dust of talc with a size of corpuscles $d = 2 \div 30$ a micron, white black and a chalk have been used. The apparatus body was filled with water on level $h_g = 0.175$ m.

Cleanliness level of an airborne dust mix was determined by a direct method [8]. On direct sections of the pipeline before and after the apparatus the mechanical sampling of an airborne dust mix was made. After determination of matching operational mode of the apparatus, gas test were taken by means of in-taking handsets. Mechanical sampling isokinetics on in-taking handsets were applied to observance replaceable tips of various diameters. Full trapping of the dust contained in taken test of an airborne dust mix, was made by an external filtering draws through mixes with the help calibrates electro-aspirator EA-55 through special analytical filters AFA-10 which were put in into filtrating cartridges. The selection time was fixed on a stopwatch, and speed—the rotameter of electro-aspirator EA-55.

Dust gas mix gained by dust injection in the flue by means of the metering screw conveyer batcher introduced on Fig. 1.6. Application of the batcher with varying productivity has given the chance to gain the set dust load on an entry in the apparatus.

FIGURE 1.6   The metering screws conveyer batcher of a dust.

The water discharge is determined by its losses on transpiration and with deleted slurry. The water drain is made in the small portions from the loading pocket supplied with a pressure lock. Gate closing implements sweeping recompression of air in the gate chamber, opening—a depressurization. Small level recession is sweepingly compensated by a top up through a connecting pipe of feeding into of a fluid. At periodic drain of the condensed slurry the water discharge is determined by consistency of slurry and averages to 10 g on 1 м³ air, and at fixed drain the charge does not exceed 100–200 g on 1 м³ air. Filling of a rotoklon with water was controlled by means of the level detector. Maintenance of a fixed level of water has essential value as its oscillations involve appreciable change as efficiency, and productivity of the device.

1.5.2   DISCUSSION OF RESULTS OF EXPERIMENT

In a rotoklon process of interacting of gas, liquid and hard phases in which result the hard phase (dust), finely divided in gas, passes in a fluid is realized. Process of hydrodynamic interacting of phases in the apparatus it is possible to disjoint sequentially proceeding stages on the following:

- fluid acquisition by a gas flow on an entry in the contact device;
- fluid subdivision by a fast-track gas flow in the contact channel;
- concretion of dispersion particles by liquid drops; and
- branch of drops of a fluid from gas in the labyrinth drip pan.

At observation through an observation port the impression is made that all channel is filled by foam and water splashes. Actually this effect caused by a retardation of a flow at an end wall, is characteristic only for a stratum, which directly is bordering on to glass. Slow-motion shot consideration allows installing a true flow pattern. It is visible that the air jet as though itself chooses the path, being aimed to be punched in the shortest way through water. Blades standing sequentially under existing conditions restrict air jet extending, forcing it to make sharper turn that, undoubtedly, favors to separation. Functionability of all dust trap depends on efficiency of acquisition of a fluid a gas flow—without fluid acquisition will not be supplied effective interacting of phases in contact channels and, hence, qualitative clearing of gas of a dust will not be attained. Thus, fluid acquisition by a gas flow at consecutive transiting of blades of an impeller is one of defined stages of hydrodynamic process in a rotoklon.

Fluid acquisition by a gas flow can be explained presence of interphase turbulence, which is advanced on an interface of gas and liquid phases. Conditions for origination of interphase turbulence is presence of a gradient of speeds of phases on boundaries, difference of viscosity of flows, and an interphase surface tension.

1.5.3   THE ESTIMATION OF EFFICIENCY OF GAS CLEANING

The quantitative assessment of efficiency of acquisition in apparatuses of shock-inertial type with inner circulation of a fluid is expedient for conducting by means of a parameter $n = L_z/L_g$, m³/m³ equal to a ratio of volumes of liquid and gas phases in contact channels and characterizing the specific charge of a fluid on gas irrigating in channels. Obviously that magnitude n will be determined, first of all, by speed of a gas flow on an entry in the contact channel. The following important parameter is fluid level on an entry in the contact channel, which can change cross-section of the channel and influence speed of gas.

$$\frac{\vartheta_g}{S_g} = \frac{\vartheta_g}{bh_k - bh_l} - \frac{\vartheta_g}{b(h_k - h_l)} \tag{7}$$

where, S_g is the cross-section of the contact channel; b is the channel width; h_k is the channel altitude; h_l is the fluid level.

Thus, for the exposition of acquisition of a fluid a gas flow in contact channels of a rotoklon it is enough to gain the following relation experimentally:

$$n = f(\vartheta_g \cdot h_l) \tag{8}$$

As it has been installed experimentally, efficiency of trapping of corpuscles of a dust in many respects depends on a size of drops of a fluid, with decrease of a size of drops the dust clearing efficiency raises. Thus, the given stage of hydrodynamic interacting of phases is rather important. For calculation of average diameter of the drops organized at transiting of blades of an impeller, the empirical relation is gained.

$$d = \frac{467 \cdot 10^3 \sqrt{\sigma}}{\vartheta_o} + 17,869 \cdot \left(\frac{\mu_l}{\sqrt{\rho_l \sigma}}\right)^{0,68} \frac{L_l}{L_r} \tag{9}$$

where ϑ_i is the relative speed of gases in the channel, m/s; σ is the factor of a surface tension of a fluid, N/m; ρ_1 is the fluid density, kg/m³; μ_l is the viscosity of a fluid,; L_l is the volume-flow of a fluid,; L_g is the volume-flow of gas,.

The offered formula allows considering also together with physical performances of phases and agency of operating conditions.

In Fig. 1.7, the design values of average diameter of the drops organized at transiting of blades of an impeller, from speed of gas in contact channels and a gas specific irrigation are introduced. At calculation values of physical properties of water were accepted at temperature 20°C: $\rho_1 =$ 998 kg/m³; $\mu_1 = 1.002 \times 10^{-3}$ N × C/m², $\varsigma = 72.86 \times 10^{-3}$ N/m.

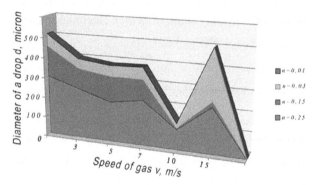

FIGURE 1.7   Computational relation of a size of drops to flow velocity and a specific irrigation.

The gained relations testify that the major operating conditions on which the average size of drops in contact channels of a rotoklon depends, speed of gas flow ϑ_r and the specific charge of a fluid on gas irrigating n are. These parameters determine hydrodynamic structure of an organized water gas flow.

Separation efficiency of gas bursts in apparatuses of shock-inertial act can be discovered only on the basis of empirical data on particular constructions of apparatuses. Methods of the calculations, found application in projection practice, are grounded on an assumption about a capability of linear approximation of relation of separation efficiency from diameter of corpuscles in is likelihood-logarithmic axes. Calculations on a likelihood method are executed under the same circuit design, as for apparatuses of dry clearing of gases [9].

Shock-inertial sedimentation of corpuscles of a dust occurs at flow of drops of a fluid by a dusty flow therefore the corpuscles possessing inertia, continue to move across the curved stream-lines of gases, the surface of drops attain and are precipitated on them.

Efficiency of shock-inertial sedimentation η_u is function of following dimensionless criterion:

$$\eta_{\grave{e}} = f\left(\frac{m_p}{\xi_c} \cdot \frac{\vartheta_p}{d_0}\right)$$

(10)

where, m_p is the mass of a precipitated corpuscle; ϑ_p is the speed of a corpuscle; ξ is the factor of resistance of driving of a corpuscle; d_0 is the diameter a midelev of cross-section of a drop.

For the spherical corpuscles which driving obeys the law the Stokes, this criterion looks like the following:

$$\frac{m_p \vartheta_p}{\xi_c d_0} = \frac{1}{18} \cdot \frac{d_r^2 \vartheta_p \rho_p C_c}{\mu_g d_0}$$

(11)

Complex $d_p^2 \vartheta_p \rho_p C_c / (18\mu_g d_0)$ is parameter (number) of the Stokes.

$$\eta_{\grave{e}} = f(Stk) = f\left(\frac{d_p^2 \vartheta_p \rho_p C_c}{18\mu_g d_0}\right)$$

(12)

Thus, efficiency of trapping of corpuscles of a dust in a rotoklon on the inertia model depends primarily on performance of a trapped dust (a size and density of trapped corpuscles) and operating conditions major of which is speed of a gas flow at transiting through blades of impellers (Fig. 1.8).

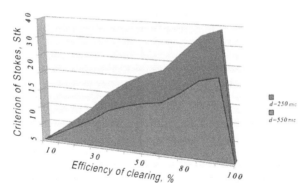

FIGURE 1.8   Relation of efficiency of clearing of gas to criterion StK.

On the basis of the observed inertia of model the method of calculation of a dust clearing efficiency in scrubbers with inner circulation of a fluid is developed.

The basis for calculation on this model is the Eq. (12). For calculation realization it is necessary to know disperse composition of a dust, density of corpuscles of a dust, viscosity of gas, speed of gas in the contact channel and the specific charge of a fluid on gas irrigating.

Calculation is conducted in the following sequence:

- by Eq. (9) determine an average size of drops D_0 in the contact channel at various operating conditions;
- by Eq. (10) count the inertia parameter of the Stokes for each fraction of a dust;
- by Eq. (11) fractional values of efficiency η for each fraction of a dust;
- general efficiency of a dust separation determine by Eq. (12), %.

The observed inertia model full enough characterizes physics of the process proceeding in contact channels of a rotoklon.

1.5.4 COMPARISON OF EXPERIMENTAL AND COMPUTATIONAL RESULTS

Analyzing the gained results of researches of general efficiency of a dust separation, it is necessary to underscore that in a starting phase of activity of a dust trap for all used in researches a dust separation high performances, components from 93.2% for carbon black to 99.8% for a talc dust are gained. Difference of general efficiency of trapping of various types of a dust originates because of their various particle size distributions on an entry in the apparatus, and also because of the various form of corpuscles, their dynamic wettability and density. The gained high values of general efficiency of a dust separation testify to correct selection of constructional and operation parameters of the studied apparatus and indicate its suitability for use in engineering of a wet dust separation.

As appears from introduced in Figs. 1.9 and 1.10 graphs, the relation of general efficiency of a dust separation to speed of a mixed gas and fluid level in the apparatus will well be agreed to design data that confirms an acceptability of the accepted assumptions.

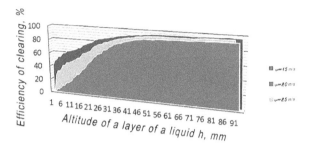

FIGURE 1.9   Relation of efficiency of clearing of gas to irrigating liquid level.

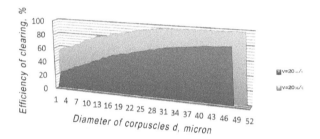

FIGURE 1.10   Dependence of efficiency of clearing of gas on a size of corpuscles and speed of gas

In Fig. 1.11, the results of researches on trapping various a dust in a rot-oklon with adjustable sinusoidal are shown. The given researches testify to a high performance of trapping of corpuscles of thin a dust with their various moistening ability. From these drawings by fractional efficiency of trapping it is obviously visible, what even for corpuscles a size less than 1 microns (which are most difficultly trapped in any types of dust traps) Installations considerably above 90%. Even for the unwettable sewed type of white black general efficiency of trapping more than 96%. Naturally, as for the given dust trap lowering of fractional efficiency of trapping at decrease of sizes of corpuscles less than 5 microns, however not such sharp, as or other types of dust traps is characteristic.

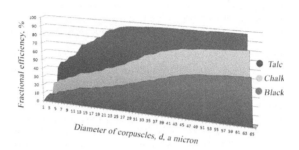

FIGURE 1.11   Fractional efficiency of clearing of corpuscles of a various dust.

1.6   CONCLUSIONS

1. The new construction of the rotoklon is developed, allowing to solve a problem of effective separation of a dust from a gas flow. In the introduced apparatus water admission to contact zones implements as a result of its circulation in the device.

2. Experimentally it is shown that fluid acquisition by a gas flow at consecutive transiting of blades of an impeller is one of defined stages of hydrodynamic process in a rotoklon.

3. Are theoretically gained and confirmed by data of immediate measurements of value of efficiency of shock-inertial sedimentation of dispersion particles in a rotoklon. The gained computational relationships, allow to size up the contribution as performances of a trapped dust (a size and density of trapped corpuscles), and operating conditions major of which is speed of a gas flow at transiting through blades of impellers.

4. Good convergence of results of scaling's on the gained relationships with the data, which are available in the technical literature and own experiments confirms an acceptability of the accepted assumptions.

The formulated leading-outs are actual for intensive operation wet-type collectors in which the basic gear of selection of corpuscles is the gear of the inertia dust separation.

KEYWORDS

- circulation of a liquid
- experimental
- gas purification
- hydrodynamic interacting of phases
- rotoklon
- scrubbers with inner circulation

REFERENCES

1. Uzhov, N. V., Valdberg, J. A., Myagkov, I. B., *Clearing of industrial gases of a dust,* Moscow: Chemistry, 1981.
2. Pirumov, I. A., *Air Dust removal,* Moscow: Engineering industry, 1974.
3. Shvydky, S. V., Ladygichev, G. M., *Clearing of gases. The directory,* Moscow: Heat power engineering, 2002.
4. V. Straus. *Industrial clearing of gases* Moscow: Chemistry, 1981.
5. Kouzov, A. P., Malgin, D. A., Skryabin, M. G., *Clearing of gases and air of a dust in the chemical industry.* St.-Petersburg: Chemistry, 1993.
6. Patent 2317845 RF, IPC, cl. B01 D47/06 *Rotoklon a controlled sinusoidal blades,* Usmanova, R. R., Zhernakov, V. S., Panov, A. K.-Publ. 27.02.2008. Bull. № 6.
7. Usmanova, R. R., Zaikov, E. G., Stoyanov, V. O., E. Klodziuska, *Research of the mechanism of shock-inertial deposition of dispersed particles from gas flow the* bulletin of the Kazan technological university №9, 203–207, 2013.
8. Kouzov, A. P., *Bases of the analysis of disperse composition industrial a dust.* Leningrad: Chemistry, 1987, 183–195.
9. Vatin, I. N., Strelets, I. K., *Air purification by means of apparatuses of type the cyclone separator.* St.-Petersburg, 2003.

CHAPTER 2

LECTURE NOTE ON QUANTUM-CHEMICAL MECHANISM AND SYNTHESIS OF SELECTED COMPOUNDS

V. A. BABKIN, V. U. DMITRIEV, G. A. SAVIN, E. S. TITOVA, and G. E. ZAIKOV

CONTENTS

2.1   INTRODUCTION

Quantum-chemical research of the mechanism of synthesis of 1-[2-(o-acetylmethyl)-3-o-acetyl-2-ethyl]-methyldichlorinephosphite for the first time is executed by classical method MNDO. This compound is received by reaction of bimolecular nucleophylic substitution S_N2, proceeding between acetyl chloride and 5-acetyloxymethyl-2-chlorine-5-ethyl-1,2,3-dioxaphosphorynan. Reaction is endothermic and has barrier character. The size of a power barrier makes 176 kDg/mol.

The mechanism of reaction acylation of bicyclophosphites by chlorine anhydride of carboxylic acids consists of three stages. The mechanism of the first stage is studied in Ref. [1]. Results of research of the second stage of this reaction are presented in the yielded work. The second stage represents interaction of acetyl chloride and 5-acetyloxymethyl-2-chlorine-5-ethyl-1,2,3-dioxaphosphorynan in a gas phase.

Now at an electronic level the mechanism of reaction of synthesis of 1-[2-(o-acetylmethyl)-3-o-acetyl-2-ethyl]-methyldichlorinephosphite   is not studied. In this connection the aim of this chapter is the quantum-chemical research of the mechanism of synthesis of this compound by quantum-chemical method MNDO.

2.2   EXPERIMENTAL

The quantum-chemical semiempirical method MNDO with optimization of geometry on all parameters by the standard gradient method, which has been built in PC GAMESS [2] has been chosen for research of the mechanism of synthesis of 1-[2-(o-acetylmethyl)-3-o-acetyl-2-ethyl]-

methyldichlorinephosphite. This method well enough reproduces power characteristics and stability of chemical compounds, including the substances containing multiple bonds [3]. Calculations were carried out in approach to the isolated molecule to a gas phase. The Program MacMolPlt was used for visual representation initial, intermediate and final models [4].

The mechanism of synthesis of studied compound was investigated by method MNDO. Initial models of components of synthesis—5-acetyloxymethyl-2-chlorine-5-ethyl-1,2,3-dioxaphosphorynan and acetyl chloride—settled down on distance 2.8–3.0Å from each other. Any interactions between compounds practically are absent on such distance. Distance R_{O9C29} has been chosen as coordinate of reaction (Fig. 2.1). This coordinate is most energetically a favorable direction of interaction of initial components. Further, optimization on all parameters of initial components at $R_{O9C29} = 2.8$Å was carried out. After optimization of value of lengths of bonds and valent corners, values E_0 (the general energy of system) and q_H-charges on atoms along coordinate of reaction R_{O9C29} were fixed and brought in Tables 2.1–2.3. The coordinate of reaction changed from 2.8Å up to 1.3Å at each step of optimization. The step on coordinate of reaction R_{O9C29} has made 0.2Å[5].

2.3   RESULTS OF CALCULATIONS

The optimized geometrical and electronic structure of initial models (a stage of coordination), an intermediate condition of system (during the moment of separation atom Cl_{32}) and a final condition of molecule of 1-[2-(o-acetylmethyl)-3-o-acetyl-2-ethyl]-methyldichlorinephosphite, are presented on Figs. 2.1–2.3. Changes of lengths of bonds, valent corners, on atoms along coordinate of reaction R_{O9C29} with step 0.2Å are shown to the general energy of all molecular system and charges in Tables 2.1–2.3 and on schedules (Figs. 2.4–2.8). Essential changes to system of initial components at steps 1–6 (R_{O9C29} changes from 2.8Å up to 1.8Å) does not occur. Their mutual orientation from each other goes at this stage. We define this stage as a stage of coordination. Simultaneous break of bond C_{29}–Cl_{32} and almost full break of bond P_{10}–O_9 occurs at II stage of interaction (6–7 steps,

R_{O9C29} change from 1.8Å up to 1.4Å). Lengths of these bonds change from 1.88Å up to 4.14Å and from 1.64Å up to 1.93 Å accordingly (Table 2.2). We define this stage, as a stage of break of bonds. Atom P_{10} of 5-acetyloxymethyl-2-chlorine-5-ethyl-1,2,3-dioxaphosphorynan is attacked by atom Cl_{32} of acetyl chloride at a stage III. Covalent bonds $P_{10\text{-}}Cl_{32}$ (2.00 Å) and $O_{9\text{-}}C_{29}$ (R_{O9C29} = 1.4Å) are formed. This stage is final. Formation of 1-[2-(o-acetylmethyl)-3-o-acetyl-2-ethyl]-methyldichlorinephosphite occurs.

Change of charges on atoms directly participating in reaction P_{10}, C_{29} and Cl_{32}, and also change E_0 along coordinate of reaction R_{O9C29} is presented on Figs. 2.4–2.8 and in Table 2.1. Reaction has barrier character (Fig. 2.4). The size of a barrier makes 176 kDg/mol. Charges on atoms P_{10} and C_{29} change according to change E_0 (Figs. 2.5 and 2.7). Charges reach the maximal values during the moment of break of bonds C_{29}–Cl_{32} and P_{10}–O_9. The Negative charge on atom Cl_{32} is inversely to change E_0. It reaches the maximal value (on the module) during the moment of break of the same bonds. We analyze behavior of atoms directly participating in reaction of synthesis of 1-[2-(o-acetylmethyl)-3-o-acetyl-2-ethyl]-methyldichlorinephosphite P_{10}, O_9, C_{29} and Cl_{32}, change of charges on these atoms and power of reaction. We do a conclusion, that the mechanism of studied synthesis represents the coordinated process with simultaneous break of bonds $P_{10\text{-}}O_9$ and $C_{29\text{-}}Cl_{32}$ and formation of new bonds $P_{10\text{-}}Cl_{32}$ and C_{29}–O_9. It is a process of nucleophylic substitution S_N2. It is similar to the mechanism of synthesis of the first stage of acidation of bicyclophosphites by chlorine anhydrides of carboxylic acids, to the mechanism of synthesis of 5-acetyloxymethyl-2-chlorine-5-ethyl-1,2,3-dioxaphosphorynan.

Thus, the mechanism of synthesis of the second stage of acidation of bicyclophosphites for the first time is studied by quantum-chemical semiempirical method MNDO. It is shown, that synthesis of this compound—result of the coordinated interactions of acetyl chloride and 5-acetyloxymethyl-2-chlorine-5-ethyl-1,2,3-dioxaphosphorynan on the mechanism of nucleophylic substitution SN2. It is positioned, that this reaction is endothermic and has barrier character. It will qualitatively be coordinated with experiment. The size of an energy barrier of studied reaction is equal 176 kDg/mol.

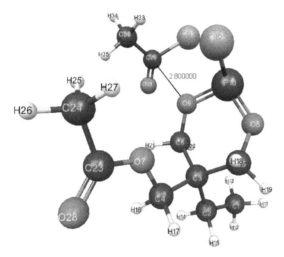

FIGURE 2.1 Initial model of interaction of 5-acetyloxymethyl-2-chlorine-5-ethyl-1,2,3-dioxaphosphorynan and acetyl chloride. $R_{O9C29} = 2.8$Å.

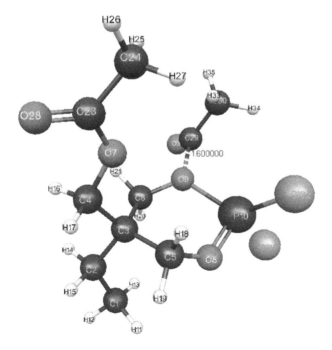

FIGURE 2.2 The Model of a stage of break of bonds (a transition state). $R_{O9C29} = 1.6$Å.

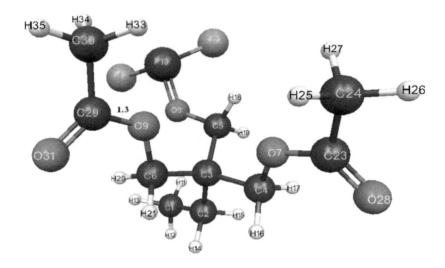

FIGURE 2.3 The Model of formation of 1-[2-(o-acetylmethyl)-3-o-acetyl-2-ethyl]-methyldichlorinephosphite. $R_{O9C29} = 1.3Å$.

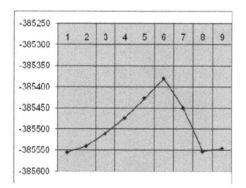

№ Step	R_{O9C29}, Å	E_0, kDg/mol
1	2,8	-385555
2	2,6	-385540
3	2,4	-385511
4	2,2	-385474
5	2,0	-385427
6	1,8	-385380
7	1,6	-385450
8	1,4	-385553
9	1,3	-385545

FIGURE 2.4 Change of the general energy of system along coordinate of reaction R_{O9C29}.

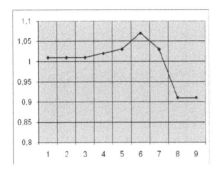

№ Step	R_{O9C29}	q, P_{10}
1	2,8	1,01
2	2,6	1,01
3	2,4	1,01
4	2,2	1,02
5	2,0	1,03
6	1,8	1,07
7	1,6	1,03
8	1,4	0,91
9	1,3	0,91

FIGURE 2.5   Change of a positive charge on atom of phosphorus P_{10} along coordinate of reaction R_{O9C29}.

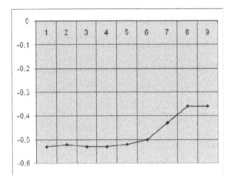

№ Step	R_{O9C29}	q
1	2,8	-0,53
2	2,6	-0,52
3	2,4	-0,53
4	2,2	-0,53
5	2,0	-0,52
6	1,8	-0,50
7	1,6	-0,43
8	1,4	-0,36
9	1,3	-0,36

FIGURE 2.6   Change of a negative charge on atom O_9 along coordinate of reaction R_{O9C29}.

№ Step	R_{O9C29}	q
1	2,8	0,34
2	2,6	0,34
3	2,4	0,35
4	2,2	0,37
5	2,0	0,39
6	1,8	0,42
7	1,6	0,38
8	1,4	0,35
9	1,3	0,35

FIGURE 2.7   Change of a positive charge on atom of carbon C_{29} along coordinate of reaction R_{O9C29}.

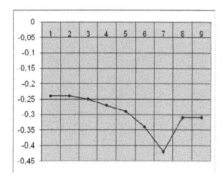

№ Step	R_{O9C29}	$E_0,$ Dg/mol
1	2,8	-0,24
2	2,6	-0,24
3	2,4	-0,25
4	2,2	-0,27
5	2,0	-0,29
6	1,8	-0,34
7	1,6	-0,42
8	1,4	-0,31
9	1 3	-0,31

FIGURE 2.8 Change of a negative charge on atom Cl_{32} along coordinate of reaction R_{O9C29}.

TABLE 2.1 Change of energy along coordinate of reaction R_{O9C29}.

$R_{O9C29,}$ Å.	$E_0,$ kDg/mol
2.8	-385555
2.6	-385540
2.4	-385511
2.2	-385474
2.0	-385427
1.8	-385380
1.6	-385450
1.4	-385553
1.3	-385545

TABLE 2.2 Change of lengths of bonds along coordinate of reaction R_{O9C29}.

$R_{C29-O9,}$ Å	R_{C1-C2}	R_{C2-C3}	R_{C3-C4}	R_{C3-C5}	R_{C3-C6}	R_{C4-O7}	R_{C5-O8}	R_{C6-O9}	R_{P10-O8}
2.8	1.53	1.57	1.58	1.58	1.58	1.40	1.39	1.39	1.58
2.6	1.53	1.57	1.58	1.58	1.58	1.40	1.39	1.39	1.58
2.4	1.53	1.57	1.58	1.57	1.58	1.40	1.39	1.40	1.58
2.2	1.53	1.57	1.58	1.57	1.58	1.40	1.39	1.40	1.58

TABLE 2.2 *(Continued)*

R_{C29-O9}, Å	R_{C1-C2}	R_{C2-C3}	R_{C3-C4}	R_{C3-C5}	R_{C3-C6}	R_{C4-O7}	R_{C5-O8}	R_{C6-O9}	R_{P10-O8}
2.0	1.53	1.57	1.58	1.57	1.58	1.40	1.39	1.40	1.57
1.8	1.53	1.57	1.58	1.57	1.58	1.40	1.38	1.42	1.57
1.6	1.53	1.57	1.58	1.58	1.58	1.40	1.38	1.41	1.58
1.4	1.53	1.57	1.59	1.57	1.57	1.40	1.39	1.40	1.57
1.3	1.53	1.57	1.59	1.57	1.57	1.40	1.39	1.41	1.57

R_{C29-O9}, Å	R_{H18-C5}	R_{H19-C5}	$R_{H;20-C6}$	R_{H21-C6}	$R_{Cl22-P10}$	R_{C23-O7}	$R_{C24-C23}$	$R_{H25-C24}$	$R_{H26-C24}$
2.8	1.12	1.11	1.12	1.12	2.04	1.36	1.52	1.10	1.10
2.6	1.12	1.11	1.12	1.12	2.04	1.36	1.52	1.10	1.10
2.4	1.12	1.11	1.12	1.12	2.04	1.36	1.52	1.10	1.10
2.2	1.12	1.11	1.12	1.12	2.04	1.36	1.52	1.10	1.10
2.0	1.12	1.11	1.12	1.12	2.03	1.36	1.52	1.10	1.10
1.8	1.12	1.11	1.12	1.11	2.03	1.36	1.52	1.10	1.10
1.6	1.12	1.12	1.12	1.12	2.04	1.36	1.52	1.10	1.10
1.4	1.12	1.12	1.12	1.12	2.01	1.36	1.52	1.10	1.10
1.3	1.12	1.12	1.11	1.12	2.01	1.36	1.52	1.10	1.10

R_{C29-O9}, Å	$R_{H27-C24}$	$R_{O28-C23}$	$R_{C30-C29}$	$R_{O31-C29}$	$R_{C132-C29}$	$R_{Cl32-P10}$	$R_{H33-C30}$	$R_{H34-C30}$	$R_{H35-C30}$
2.8	1.10	1.22	1.51	1.20	1.80	3.54	1.10	1.11	1.10
2.6	1.10	1.22	1.51	1.20	1.81	3.34	1.10	1.11	1.10
2.4	1.10	1.22	1.51	1.20	1.81	3.16	1.10	1.11	1.10
2.2	1.10	1.22	1.52	1.20	1.82	2.99	1.10	1.11	1.10
2.0	1.10	1.22	1.52	1.21	1.84	2.83	1.10	1.11	1.10
1.8	1.10	1.22	1.53	1.21	1.88	2.67	1.10	1.11	1.10
1.6	1.10	1.22	1.50	1.20	4.14	2.03	1.10	1.10	1.11
1.4	1.10	1.22	1.52	1.22	5.76	2.00	1.10	1.10	1.10
1.3	1.10	1.22	1.53	1.23	5.75	2.00	1.10	1.10	1.10

TABLE 2.3   Change of charges of system along coordinate of reaction R_{O9C29}

$R_{O9C29,}$ Å	2.8	2.6	2.4	2.2	2.0	1.8	1.6	1.4	1.3
C1	0.02	0.02	0.02	0.02	0.02	0.02	0.03	0.03	0.03
C2	0.00	0.00	0.00	0.00	0.00	0.00	0.00	−0.01	−0.01
C3	−0.16	−0.16	−0.16	−0.16	−0.16	−0.15	−0.13	−0.09	−0.09
C4	0.21	0.21	0.20	0.21	0.20	0.20	0.20	0.20	0.20
C5	0.21	0.21	0.21	0.21	0.21	0.22	0.23	0.25	0.25
C6	0.22	0.22	0.22	0.21	0.21	0.20	0.19	0.21	0.22
O7	−0.35	−0.35	−0.35	−0.35	−0.35	−0.35	−0.35	−0.34	−0.34
O8	−0.51	−0.51	−0.51	−0.51	−0.51	−0.52	−0.50	−0.50	−0.51
O9	−0.53	−0.52	−0.53	−0.53	−0.52	−0.50	−0.43	−0.36	−0.32
P10	1.01	1.01	1.01	1.02	1.03	1.07	1.03	0.91	0.91
H11	0.00	0.00	0.00	0.00	0.00	0.00	0.01	0.01	0.01
H12	0.00	0.00	0.00	0.00	0.00	0.00	0.00	0.00	0.00
H13	0.00	0.00	0.00	0.00	0.00	0.00	0.01	0.01	0.01
H14	0.01	0.01	0.01	0.01	0.01	0.01	0.01	0.01	0.01
H15	0.01	0.01	0.01	0.01	0.01	0.01	0.01	0.01	0.01
H16	0.01	0.01	0.01	0.01	0.01	0.01	0.01	0.01	0.01
H17	0.01	0.01	0.01	0.01	0.01	0.02	0.02	0.01	0.01
H18	0.02	0.01	0.02	0.02	0.02	0.02	0.01	0.01	0.01
H19	0.03	0.03	0.04	0.04	0.04	0.04	0.03	0.01	0.01
H20	0.01	0.01	0.02	0.02	0.02	0.03	0.02	0.01	0.02
H21	0.04	0.03	0.03	0.03	0.03	0.04	0.01	0.01	0.02
Cl22	−0.42	−0.41	−0.41	−0.41	−0.41	−0.40	−0.48	−0.36	−0.36
C23	0.35	0.35	0.35	0.35	0.35	0.35	0.35	0.35	0.35
C24	0.05	0.05	0.05	0.05	0.05	0.05	0.05	0.05	0.05
H25	0.03	0.03	0.03	0.03	0.03	0.03	0.02	0.03	0.03
H26	0.03	0.03	0.03	0.03	0.03	0.03	0.03	0.03	0.03
H27	0.04	0.04	0.04	0.04	0.04	0.04	0.03	0.03	0.03

TABLE 2.3 *(Continued)*

R_{O9C29}, Å	2.8	2.6	2.4	2.2	2.0	1.8	1.6	1.4	1.3
O28	−0.35	−0.35	−0.35	−0.35	−0.35	−0.35	−0.35	−0.35	−0.35
C29	0.34	0.34	0.35	0.37	0.39	0.42	0.38	0.35	0.35
C30	0.02	0.02	0.03	0.03	0.03	0.03	0.05	0.05	0.05
O31	−0.23	−0.24	−0.24	−0.25	−0.28	−0.32	−0.21	−0.33	−0.38
Cl32	−0.24	−0.24	−0.25	−0.27	−0.29	−0.34	−0.42	−0.31	−0.31
H33	0.05	0.05	0.05	0.05	0.04	0.04	0.05	0.03	0.02
H34	0.03	0.03	0.03	0.03	0.03	0.04	0.06	0.03	0.03
H35	0.03	0.03	0.03	0.03	0.02	0.02	0.04	0.03	0.02

TABLE.2.4   Change of valence corners along coordinate of reaction R_{O9C29}.

Coordinate of reaction. R_{O9C29}, Å.	2.8	2.6	2.4	2.2	2.0	1.8	1.6	1.4	1.3
C(3)C(2)C(1)	119	120	120	120	120	120	120	120	120
C(4)C(3)C(2)	105	105	104	104	104	104	104	104	104
C(5)C(3)C(2)	110	110	110	110	110	110	110	111	111
C(6)C(3)C(2)	108	108	108	108	108	108	108	108	108
O(7)C(4)C(3)	110	110	110	110	111	111	111	111	111
O(8)C(5)C(3)	113	113	113	112	112	112	113	112	111
O(9)C(6)C(3)	115	116	117	117	117	117	113	111	111
P(10)O(8)C(5)	127	128	128	128	128	129	125	129	129
H(11)C(1)C(2)	112	112	112	112	112	112	112	112	112
H(12)C(1)H(11)	107	107	107	107	107	107	107	106	106
H(13)C(1)C(12)	107	107	107	107	106	106	107	106	106
H(14)C(2)C(1)	107	107	107	107	107	107	107	106	106
H(15)C(2)H(14)	105	105	105	105	105	105	105	105	105
H(16)C(4)O(7)	110	110	110	110	110	110	109	110	109
H(17)C(4)H(16)	106	106	106	106	106	106	106	106	106

TABLE 2.4   *(Continued)*

Coordinate of reaction. R_{O9C29}, Å.	2.8	2.6	2.4	2.2	2.0	1.8	1.6	1.4	1.3
H(18)C(5)O(8)	109	109	109	109	109	109	109	109	109
H(19)C(5)H(18)	104	104	104	104	104	104	105	106	106
H(20)C(6)O(9)	107	107	107	107	107	107	108	109	108
H(21)C(6)H(20)	105	105	105	105	106	106	105	106	106
Cl(22)P(10)O(8)	105	105	104	104	104	104	103	105	105
C(23)O(7)C(4)	125	124	124	124	124	124	124	124	124
C(24)C(23)O(7)	113	113	113	113	113	113	113	113	113
H(25)C(24)C(23)	110	110	110	110	110	111	110	110	110
H(26)C(24)H(25)	108	108	108	108	108	108	108	108	108
H(27)C(24)H(25)	108	108	108	108	108	108	108	108	108
O(28)C(23)O(7)	119	119	119	119	119	119	119	120	120
C(29)O(9)C(6)	115	116	117	118	118	118	118	124	126
C(30)C(29)O(9)	97	97	97	98	100	104	112	112	114
O(31)C(29)O(30)	128	128	128	128	127	126	135	127	123
Cl(32)C(29)O(9)	93	93	94	94	95	95	54	53	53
H(33)C(30)C(29)	111	112	112	112	113	113	112	111	109
H(34)C(30)H(32)	97	99	97	97	95	93	60	41	84
H(35)C(30)H(34)	108	109	108	108	108	108	108	108	107

KEYWORDS

- 1-[2-(o-acetylmethyl)-3-o-acetyl-2-ethyl]-methyldichlorinephosphite
- 5-acetyloxymethyl-2-chlorine-5-ethyl-1,2,3-dioxaphosphorynan
- acetyl chloride
- method MNDO
- quantum-chemical research
- the mechanism of synthesis

REFERENCES

1. Babkin, A. V., Dmitriev, U. V., Savin, A. G., Zaikov, E. G., Estimation of acid force of components of synthesis of 5-acetyloxymethyl-2-chlorine-5-ethyl-1,2,3-dioxaphos-phorynan. Moscow, *Encyclopedia of the engineer-chemist.* 13, 11–13 (2009).
2. Schmidt, W. M., Baldrosge, K. K., Elbert, A. J., Gordon, S. M., Enseh, H. J., S. Koseki, N. Matsvnaga, Nguyen, A. K., Su, J. S., et al., *J. Computer Chem.* 14, 1347–1363 (1993).
3. Clark, T. *The Computer chemistry.* M.: World, 1990, 383 p. Quantum-chemical research of the mechanism of synthesis.
4. Bode, B. M., Gordon, M. S. *J. Mol. Graphics Mod.* 16, 1998, 133–138.
5. Babkin, A. V., Rachimov, I. A., Titova, S. E., Fedunov R. G., Reshetnikov, A. R., Belousova, S. V., Zaikov, E. G., Quontum-chemical researches of the mechanism of synthesis 2-methil(benzil)-tio-4-methil(benzil)oxipyrimidine. Izhevsk, The chemical physics and mesoscopy. 9, 263–276.

CHAPTER 3

TRANSPORT PROPERTIES OF FILMS OF CHITOSAN—AMIKACIN

A. S. SHURSHINA, E. I. KULISH, and S. V. KOLESOV

CONTENTS

3.1   INTRODUCTION

Transport properties of films on a basis chitosan and medicinal substance are investigated. Sorption and diffusive properties of films are studied. Diffusion coefficients are calculated. Kinetic curves of release of the amikacin, having abnormal character are shown. The analysis of the obtained data showed that a reason for rejection of regularities of process of transport of medicinal substance from chitosan films from the classical Fikovsky mechanism are structural changes in a polymer matrix, including owing to its chemical modification at interaction with medicinal substance

Systems with controlled transport of medicines are the extremely demanded [1, 2]. Research of regularities of processes of diffusion of water and medicinal substance in polymer films and opportunities of control of release of medicines became the purpose of this work. As a matrix for the immobilization of drugs used naturally occurring polysaccharide chitosan, which has a number of valuable properties: nontoxicity, biocompatibility, high physiological activity [3], as well as a drug used aminoglycoside antibiotic series—amikacin, actively applied in the treatment of pyogenic infections of the skin and soft tissue. [4]

3.2   EXPERIMENTALS

The object of investigation was a chitosan (ChT) specimen produced by the company "Bioprogress" (Russia) and obtained by acetic deacetylation of crab chitin (degree of deacetylation ~84%) with M_{sd}=334,000. As the medicinal substance (MS) used an antibiotic amikacin (AM)—quadribasic aminoglycoside, used in the form of salts − sulfate (AMS) and chloride (AMCh). Chemical formulas of objects of research and their symbols used in the text, are given in Table 3.1.

TABLE 3.1 Formulas research objects and symbols used in the reaction schemes.

Formula object of study	Symbol

Chitosan acetate monomer unit

$\sim\sim\sim NH_3{}^+CH_3COO^-$

Amikacin   sulfate

Amikacin   chloride

ChT films were obtained by means of casting of the polymer solution in 1% acetic acid onto the glass surface with the formation of chitosan acetate (ChTA). Aqueous antibiotic solution was added to the ChT solution immediately before films formation. The content of the medicinal preparation in the films was 0.01, 0.05 and 0.1 mol/mol ChT. The film thickness in all experiments was maintained constant and equal 100 microns. To study the release kinetics of MS the sample was placed in a cell with distilled

water. Stand out in the aqueous phase AM recorded spectrophotometri-
cally at a wavelength of 267 nm, corresponding to the maximum absorption in the UV spectrum of MS. Quantity of AM released from the film at
time t (G_s) was estimated from the calibration curve. The establishment of
a constant concentration in the solution of MS G_∞ is the time to equilibrium. MS mass fraction α, available for diffusion, assessed as the quantity
of films released from the antibiotic to its total amount entered in the film.

Studying of interaction MS with ChT was carried out by the methods of IR- and UV-spectroscopy. IR-spectrums of samples wrote down on
spectrometer "Shimadzu" (the tablets KBr, films) in the field of 700–3600
cm^{-1}. UV-spectrums of all samples removed in quartz ditches thick of 1
cm concerning water on spectrophotometer "Specord M-40" in the field of
220–350 nanometers.

With the aim of determining the amount of medicinal preparation held
by the polymer matrix β there was carried out the synthesis of adducts of
the ChT-antibiotic interaction in acetic acid solution. The synthesized adducts were isolated by double re-precipitation of the reaction solution in
NaOH solution with the following washing of precipitated complex residue with isopropyl alcohol. Then the residue was dried in vacuum up to
constant mass. The amount of preparation strongly held by chitosan matrix
was determined according to the data of the element analysis on the analyzer EUKOEA-3000 and UF-spectrophotometrically.

The relative amount of water m_t absorbed by a film sample of ChT,
determined by an exicator method, maintaining film samples in vapors of
water before saturation, and calculated on a formula: $m_t = (\Delta m)/m_0$, where
m_0 is the initial mass of ChT in a film, Δm is the weight the absorbed film
of water by the time of t time.

Isothermal annealing of film samples carried out at a temperature 120°C
during fixed time. Structure of a surface of films estimated by method of laser
scanning microscopy on device LSM-5-Exciter (Carl Zeiss, Germany).

3.3   RESULTS AND DISCUSSION

It is well known that release of MS from polymer systems proceeds as
diffusive process [5–7]. However, a necessary condition of diffusive trans-

port of MS from a polymer matrix is its swelling in water, that is, effective diffusion of water in a polymer matrix. Diffusing in a polymer matrix, water molecules, possessing it is considerable bigger mobility in comparison with high-molecular substance, penetrate in a polymer material, separating apart chains and increasing the free volume of a sample. The main mechanisms in water transport in polymer films are simple diffusion and the relaxation phenomena in swelling polymer. If transfer is caused mainly mentioned processes, the kinetics of swelling of a film is described by the following equation [8]:

$$m/m_\infty = kt^n, \tag{1}$$

where, m_∞ is the relative amount of water in equilibrium swelling film sample, k is the a constant connected with parameters of interaction polymer - diffuse substance, n is an indicator characterizing the mechanism of transfer of substance. If transport of substance is carried out on the diffusive mechanism, the indicator of n has to be close to 0.5. If transfer of substance is limited by the relaxation phenomena, $n > 0.5$.

The parameter n determined for a film of pure ChT is equal 0,63 (i.e., > 0.5) that is characteristic for the polymers, being lower than vitrification temperature [9]. This fact is connected with slowness of relaxation processes in glassy polymers. Values of equilibrium sorption of water and indicator n defined for film samples, passed isothermal annealing (a relaxation of nonequilibrium conformations of chains with reduction of free volume), are presented in Table 3.2.

TABLE 3.2   Parameters of swelling of chitosan films in water vapor.

Composition of the film	The concentration of MS in the film, mol/mol ChT	Annealing time, min	$D_s^a *10^{11}$, cm²/sec	$D_s^6 *10^{11}$, cm²/sec	n	Q_y, g/g ChT
		15	37.2	37.0	0.50	2.50
ChT		30	36.3	36.0	0.48	2.48
		60	15.3	14.1	0.44	2.47
		120	12.2	13.0	0.43	2.46

TABLE 3.2 *(Continued)*

Composition of the film	The concentration of MS in the film, mol/mol ChT	Annealing time, min	$D_s^a *10^{11}$, cm²/sec	$D_s^6 *10^{11}$, cm²/sec	n	Q_v, g/g ChT
	0.01	30	8.5	4.7	0.42	1.84
		60	7.0	4.5	0.39	1.56
ChT-AMCh		120	5.8	4.2	0.37	1.42
	0.05	30	5.3	3.2	0.34	1.85
	0.1	30	4.4	3.1	0.32	1.48
	0.01	30	6.9	2.9	0.34	1.58
		60	4.0	2.8	0.27	1.46
ChT-AMS		120	2.6	2.3	0.25	1.31
	0.05	30	6.1	2.2	0.30	1.66
	0.1	30	2.7	1.9	0.27	1.07

Apparently from Table 3.2 data, carrying out isothermal annealing leads to that values of an indicator n decrease. Thus, if annealing was carried out during small time (15–30 minutes), the value *n* determined for pure ChT is close to 0.5. It indicates that transfer of water is limited by diffusion, and it is evidence that ChT in heat films is in conformational relaxed condition. In process of increase time of heating till 60–120 minutes, values of an indicator n continue to decrease that, most likely, reflects process of further restructuring of the polymer matrix, occurring in the course of film heating. That processes of isothermal annealing of ChT at temperatures ≥ 100°C are accompanied by course of a number of chemical transformations, was repeatedly noted in literature [10,11]. In particular, it is revealed that besides acylation reaction, there is the partial destruction of polymer increasing the maintenance of terminal aldehyde groups which reacting with amino groups, sew ChT macromolecules at the expense of formation of azomethine connections. In Ref. [12], the fact of cross-linking in the HTZ during isothermal annealing was confirmed by the study of the spin-lattice relaxation. In the values of equilibrium sorption isothermal annealing, however, actually no clue, probably owing to the low density of cross-links.

A similar result—reduction indicator n is achieved when incorporated into a polymer matrix MS. As the data in Table 3.2, the larger MS entered into the film, the slower and less absorb water ChT. During isothermal annealing medicinal films effect enhanced. Such deviations from the laws of simple diffusion (Fick's law) and others researches have observed, explaining their strong interaction polymer with MS [13].

In the aqueous environment from ChT film with antibiotic towards to the water flow moving to volume of the chitosan, from a polymeric film LV stream is directed to water.

In Fig. 3.1, typical experimental curves of an exit of AM from chitosan films with different contents of MS are presented. All the kinetic curves are located on obviously expressed limit corresponding to an equilibrium exit of MS (G_∞).

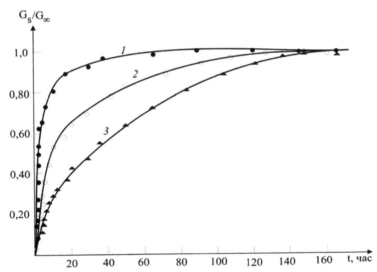

FIGURE 3.1 Kinetic curves of the release of the MS from film systems ChT-AMCh with the molar ratio of 1:0.01 (1), 1:0.05 (2) and 1:0.1 (3). Isothermal annealing time – 30 minutes.

Mathematical description of the desorption of low molecular weight component from the plate (film) with a constant diffusion coefficient is in detail considered in Ref. [14], where the original differential equation adopted formulation of the second Fick's law:

$$\frac{\partial c_s}{\partial t} = D_s \frac{\partial^2 c_s}{\partial x^2} \tag{1}$$

The solution of this equation disintegrates to two cases: for big ($G_s/G_\infty >$ 0.5) and small ($G_s/G_\infty \leq 0.5$) experiment times:

On condition $G_s/G_\infty \leq 0.5$ $\qquad$ $G_s/G_\infty = [16\, D_s t/\pi L^2]^{0.5}$ $\qquad$ (2a),

On condition $Gs/G_\infty > 0.5$ $\qquad$ $G_s/G_\infty = 1 - [8/\pi^2 \exp(-\pi^2 D_s t/L^2]$ $\qquad$ (2b),

where $G_s(t)$ is the concentration of the desorbed substance at time t and $G_\infty - Gs$ value at $t \to \infty$, L is the thickness of the film sample.

In case of transfer MS with constant diffusion coefficient the values calculated as on initial (condition $G_s/G_\infty \leq 0.5$), and at the final stage of diffusion (condition $G_s/G_\infty > 0.5$), must be equal. The equality $D_s^a = D_s^b$ indicates the absence of any complications in the diffusion system polymer—low-molecular substance [15]. However, apparently from the data presented in Table 3.3, for all analyzed cases, value of diffusion coefficients calculated at an initial and final stage of diffusion don't coincide.

TABLE 3.3 Desorption parameters AMS and AMCh in films based on ChT.

Composition of the film	The concentration of MS in the film, mol/mol ChT	Annealing time, min	$D_s^a*10^{11}$, cm²/sec	$D_s^b*10^{11}$, cm²/sec	n	α
ChT-AMCh	1:0.01	30	81.1	3.3	0.36	0.95
		60	76.9	3.0	0.33	0.92
		120	37.0	2.9	0.25	0.88
	1:0.05	30	27.2	2.5	0.29	0.91
	1:0.1	30	25.9	2.2	0.21	0.83
ChT-AMS	1:0.01	30	24.6	1.9	0.30	0.94
		60	23.0	1.8	0.22	0.90
		120	21.1	1.4	0.20	0.85
	1:0.05	30	24.0	1.8	0.18	0.80
	1:0.1	30	23.3	1.5	0.16	0.70

Note that in the process of water sorption similar regularities (Table 3.2) were observed. This indicates to a deviation of the diffusion of the classical type and to suggest the so-called pseudo-normal mechanism of diffusion of MS from a chitosan matrix.

About pseudo-normal type diffusion MS also shows kinetic curves, constructed in coordinates $G_s/G_\infty - t^{1/2}$ (Fig. 3.2). In the case of simple diffusion, the dependence of the release of the MS from film samples in coordinates $G_s/G_\infty - t^{1/2}$ would have to be straightened at all times of experiment. However, as can be seen from Fig. 3.2, a linear plot is observed only in the region $G_s/G_\infty < 0.5$, after which the rate of release of the antibiotic significantly decreases.

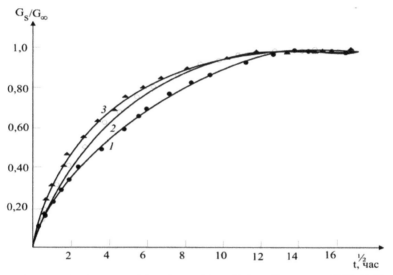

FIGURE 3.2   Kinetic curves of the release of the MS from film systems HTZ-AMS with a molar concentration of 0.01 (1) 0.05 (2) and 0.1 (3). Isothermal annealing time – 30 minutes.

At diffusion MS from films, also as well as in case of sorption by films ChT−AM of water vapor, have anomalously low values of the parameter n, estimated in this case from the slope in the coordinates $\lg(G_s/G_\infty) - \lg t$. Increase the concentration of MS and time of isothermal annealing, as well as in the process of sorption of water vapor films, accompanied by an additional decrease in the parameter n. Symbatically index n also changes magnitude α. Moreover, the relationship between the parameters of water sorption films ChT-AM (values of diffusion coefficient, an indicator n and values of equilibrium sorption Q_∞) and the corresponding parameters of the diffusion of MS from the polymer film on condition $G_s/G_\infty \leq 0.5$ is shown.

All known types of anomalies of diffusion, can be described within relaxation model [16]. Unlike the Fikovsky diffusion assuming instant establishment and changes in the surface concentration of sorbate, the relaxation model assumes change in the concentration in the surface layer on the first-order equation [17]. One of the main reasons causing change of boundary conditions is called nonequilibrium of the structural-morphological organization of a polymer matrix [16]. A possible cause of this could be the interaction between the polymer and LV.

Abnormally low values *n* in Ref. [18] were explained with presence tightly linked structure of amorphous-crystalline matrix. In this case, is believed, the effective diffusion coefficient in process of penetration into the volume of a sample can be reduced due to steric constraints that force diffusant bypass crystalline regions and diffuse to the amorphous mass of high-density cross-linking. As ChT belongs to the amorphous-crystalline polymers, it would be possible to explain low values n observed in our case similarly. However, according to Table 3.2 in the case of films of individual ChT such anomalies are not observed. Thus, the effect of substantially reducing n is associated with the interaction between HTZ and LV.

In the volume of the polymer matrix MS can be in different states. Part of the drug may be linked to the macromolecular chain through any chemical bonds, on the other hand, some of it may be in the free volume in the form of physical filler. In the latter case, it can cause a certain structural organization of the polymer matrix. That antibiotic AM influences on the structure and morphology of the films ChT, indicate data of the laser scanning microscopy. As seen from the electron microscope images of Fig. 3.3, the initial films ChT visible surface strong interference caused by its heterogeneity. With the introduction of the film AM, the interference surface is significantly reduced.

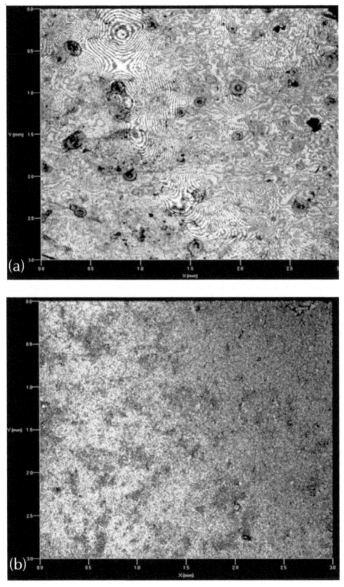

FIGURE 3.3   Micrograph of the surface (in contact with air) film individual ChT (a) and film ChT-AMS (b).

IR-and UV-spectroscopy data indicate to taking place interaction be-tween AM and ChT. It may be noted, for example, a significant change in the ratio between the intensity of the bands ChT corresponding hydroxyl and nitrogen-containing groups, before and after the interaction with the antibiotic (Table 3.4).

TABLE 3.4 The value of the intensity ratio of the absorption bands of some of the data of the IR spectra.

Sample	I_{1640}/I_{2900}	I_{1590}/I_{2900}	I_{1458}/I_{2900}
ChT	0.57	0.51	0.72
AM	0.61	0.62	0.7
ChT-AM complex derived from 1% acetic acid	0.48	0.53	0.59

Binding energy in the adduct reaction ChT-antibiotic, estimated by the shift in the UV spectra of the order of 10 kJ/mol, which allows to tell about connection ChT-antibiotic by hydrogen bonds.

Thus, AM may interact with ChT by forming hydrogen bonds. However, interpretation of the data on the diffusion is much more important that AM can form chains linking ChT by salt formation.

Exchange interactions between ChT and AMS may occur under the scheme:

$$2 \quad \underset{\underset{NH_3^+}{\overset{NH_3^+}{\big|}}}{\underset{O^-}{\overset{O^-}{S}}} \quad + \quad \begin{array}{c} CH_3COO^-NH_3^+ \\ CH_3COO^-NH_3^+ \end{array} AM \begin{array}{c} NH_3^+CH_3COO^- \\ NH_3^+CH_3COO^- \end{array}$$

Due dibasic sulfuric acid, it is possible to suggest the formation of two types of salts, providing stapling ChT macromolecules with the loss of its solubility. Firstly, the water-insoluble "double" salt—sulfate ChT-AMS, secondly, the salt mixture—insoluble in water ChT sulfate and soluble AM acetate.

If to take the AM in the form of chloride, an exchange reaction between ChT acetate and AM chloride reduces the formation of dissociated soluble salts. Accordingly, the reaction product in this case will consist of the H-complex ChT-AM.

Data on a share of antibiotic related to polymer adducts (β), obtained in solutions of acetic acid, are presented in Table 3.5.

TABLE 3.5 Mass fraction of the antibiotic β, defined in reaction adducts obtained from 1% acetic acid.

Used antibiotic	The concentration of MS in the film, mol/mol ChT	β
AMS	1.00	0.72
	0.10	0.33
	0.05	0.21
	0.01	0.07
AMCh	1.00	0.37
	0.10	0.20
	0.05	0.08
	0.01	0.04

As seen in Table 3.5, from the fact that AMC is able to "sew" chitosan chain is significantly more closely associated with macromolecules MS than for AMX.

Formation of chemical compounds of MS with ChT is probably the reason for the observed anomalies—reducing the rate of release of MS from film caused by simple diffusion, as well as the reduction of the share allocated to the drug (α). Indeed, the proportion of MS found in adducts of reaction correlates with the share of the antibiotic is not capable of participating in the diffusion process, and with the index n, reflecting the diffusion mechanism (Fig. 3.4).

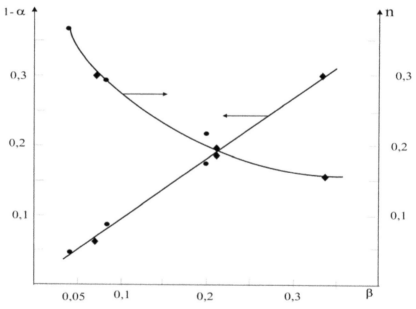

FIGURE 3.4 The fraction of amikacin, not involved in the diffusion of amikacin share determined in the adduct reaction (• – system ChT-AMS; • – ChT-AMCh).

Thus, structural changes in the polymer matrix, including as a result of its chemical modification of the interaction with the drug substance, cause deviations regularities of transport MS of chitosan films from classic Fikovsky mechanism. Mild chemical modification, for example, by cross-linking macromolecules salt formation, not affecting the chemical structure of the drug, is a possible area of control of the transport properties of medicinal chitosan films.

KEYWORDS

- **Fikovsky mechanism**
- **medicinal chitosan films**
- **polymer matrix**
- **salt formation**

REFERENCES

1. Shtilman, M. I. Polimeryi medicobiologicheskogo naznacheniya. M.: Akademkniga, 2006, 58.
2. Plate, N. A., Vasilev, A. E. Fiziologicheski aktivnyie polimeryi. M.: Himiya, 1986, 152.
3. Skryabin, K. G., Vihoreva, G. A., Varlamov, V. P. Hitin I hitozan. Poluchenie, svoistva i primenenie. M.: Nauka, 2002, 365.
4. Mashkovskii, M. D. Lekarstvennyie sredstva. Harkov: Torsing, 1997, T.2. 278.
5. Ainaoui, A., Verganaud, J. M., Comput.Theor.polym.Sci, 2000, V. 10. № 2. 383.
6. Kwon, J.-H., Wuethrich, T., Mayer, P., Escher, B. I., Chemosphere, 2009, V. 76. 83.
7. Martinelli, A., D¢Ilario, L., Francolini, I., Piozzi, A., Int.J.Pharm, 2011, V. 407. № 1–2. 197.
8. Hall, P. J., Thomas, K. M., Marsh, Fuel, 1992,V.71. №11. 1271.
9. Chalyih, A. E. Diffuziya v polimernyih sistemah. M.: Himiya, 1987, 136.
10. Zotkin, M. A., Vihoreva, G. A., Ageev, E. P. etc., Himicheskaya tehnologiya, 2004, № 9. 15.
11. Ageev, E. P., Vihoreva, G. A., Zotkin, M. A. etc., Vyisokomolekulyarnyie soedineniya, 2004, T.46. № 12, 2035.
12. Smotrina, T. V., Butlerovskie soobscheniya, 2012, T.29. №2. 98–101.
13. Singh, B., Chauhan, N., Acta Biomaterialia, 2008, V.4. № 1. 1244.
14. Crank, J. The Mathematics of Diffusion. Oxford: Clarendon Press, 1975, 46.
15. Ukhatskaya, E. V., Kurkov, S. V., Matthews, S. E. etc., Int. J. Pharm, 2010, V. 402. № 1–2. 10.
16. Malkin, A. Ya., Chalyih, A. E. Diffuziya i vyazkost polimerov. Metodyi izmereniya. M.: Himiya, 1979, 304.
17. Pomerancev, A. L. Metodyi nelineinogo regressionnogo analiza dlya modelirovaniya kinetiki himicheskih I fizicheskih processov. Dis. D. fiz.-mat. nauk. M.: MGU, 2003.
18. Kuznecov, P. N., Kuznecova, L. I., Kolesnikova, S. M., Himiya v interesah ustoichivogo razvitiya, 2010, №18. 283–298.

CHAPTER 4

RESEARCH AND CALCULATION OF OPERATING CONDITIONS OF CLEARING OF GAS IN A ROTOKLON

R. R. USMANOVA and G. E. ZAIKOV

CONTENTS

4.1   INTRODUCTION

The analysis of hydraulic losses in blade impellers is given. Computer simulation has been applied to studying of hydrodynamic characteristics in the apparatus in program ANSYS-14 CFX. It is shown that it is expedient to apply aerodynamically rational roll forming of shovels of an impeller to hydraulic resistance decrease. Effect of viscosity of an irrigating liquid on gas cleaning process is investigated. The conducted researches give the grounds for acknowledging of a hypothesis on existence of such boundary concentration of suspension at which excess the apparatus overall performance decreases.

Now necessity for creation of such wet-type collectors which would work with the low charge of an irrigating liquid has matured and combined the basic virtues of modern means of clearing of gases: simplicity and compactness, a high performance, a capability of control of processes of a dust separation and optimization of regimes.

To the greatest degree modern demands to the device and activity of apparatuses of clearing of industrial gases there match wet-type collectors with inner circulation the fluids gaining now more and more a wide circulation in systems of a gas cleaning in Russia and abroad.

Known constructions of dust traps have a broad band of regulating of one or both operating conditions of clearing of gas. It allows to expand considerably area of their application and to use the given apparatuses not only for trapping of a coarse dust of auxiliaries, but also for clearing of spent gases of the basic industrial assemblies of a mesh disperse dust.

At the expense of gas speed control in contact channels and optimization of the specific charge of a fluid on gas irrigating it is possible to ensure functioning of the given dust traps in an optimum regime, that is, to conduct clearing of gases of a particular aspect of a dust with peak efficiency at minimum power inputs.

Installation a rotoklon with adjustable sinusoidal lobes (Fig. 4.1) has been chosen because in it in a turbulent regime all gripping mechanisms of parts of a dust by water, except the filtration are realized. In it trapping gears, characteristic for the wet cyclone separator as lobes have some hemispherical turns are realized. The gripping mechanism of corpuscles of a dust is realized by liquid drops which are organized in a great many in impellers at speeds of driving of gas of 25 m/s and more also, capture parts of a dust and, being propelled with high speeds, are precipitated on walls of lobes. On an exit from lobes the fluid is rejected in the form of a tangential film, which is dispersed by an air-out in the form of a field of drops, and a turbulent stratum of foam as the foamy gripping mechanism of corpuscles also is successfully realized in the given installation. The stratum of foam unlike foamy apparatuses exists only at air high speeds, it is unstable and is instantly destroyed at air stopping delivery.

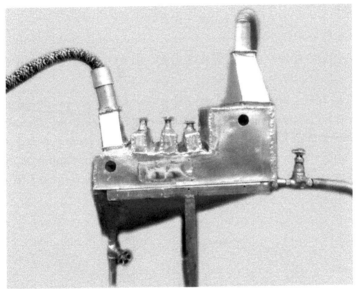

FIGURE 4.1   Installation a rotoklon with adjustable sinusoidal lobes [1].

Implementation of almost all gears of a wet dust separation and higher stability in-process in comparison with an ordinary rotoklon at rather broad band of change of speeds of gas also have predetermined sampling and working out of the examined apparatus.

By dust trap working out results of the theoretical analysis, which have confirmed are considered that the major operating condition defining hydrodynamics and efficiency of process of clearing, speed of gas in a free cross-sectional area of contact devices is. Other operating condition which should be optimized according to first, the specific charge of a fluid on gas irrigating in contact blade impellers.

4.2   THE PURPOSE AND RESEARCH PROBLEMS

- To make the analysis of hydraulic losses in blade impellers and to state a comparative estimation of various constructions of contact channels of an impeller by efficiency of security by them of hydro-dynamic interacting of phases.
- Definition of boundary densities of suspension various a dust after which excess general efficiency of a dust separation is reduced.
- Definition of the maximum extent of circulation of an irrigating liquid.

4.3   AERODYNAMIC PROFILING OF BLADES OF THE IMPELLER

Giving to blade impellers sinusoidal a profile allows to eliminate flow brake-off on crimps. Thus there is a flow of an input section of a profile of blades with large fixed speed and increase in skips from profiled parts of blades with which account it is possible to forecast negligible increase in efficiency of clearing of gas. The calculation circuit design profiled blades is figured on Fig. 4.2.

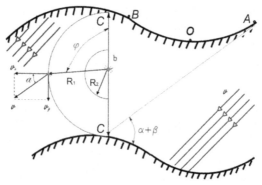

FIGURE 4.2   the Computational circuit design profiled impeller blades.

In an input part of a dust-laden gas flow free boundary lines of a stream are substituted profiled by channels of blades of an impeller. In a blow point about *a* relative wind it is atomized and flows round curvilinear surface

AV, thus speed attains the maximum value ϑ_{max} and it is saved by a constant on length of all section. In cross-section about a flow it is possible to observe as uniform, moving with a speed ϑ, and blade impellers have final width, brake-off of a flow from impeller blades does not occur. Flow implements lost-free a total pressure to critical cross-section S-S after which static pressure in a flow starts to drop. Hydrodynamic it is expedient to profile blade impellers taking into account various values of speeds within slotted channels. Thus, the dust-laden gas flow can be turned on any necessary angle with the least hydraulic losses.

As have shown results of experiment, angle change α from 30° to 45° leads to increase in a flow area of flow R_2–R_1 at veering of driving of gas and to increase *b* after its turn. The flow area *b* is increased on the average in 1.55 times. Thus speed of gas at transiting through blade impellers is reduced that involves decrease of a tangential component ϑ_y speeds of dispersion particles. The centrifugal force of inertia responsible for branch of corpuscles from a gas flow and defining an increment of the radial component ϑ_h of speed of corpuscles, thus also is reduced. Expansion of stream R_2–R_1 promotes increase in a mechanical trajectory of corpuscles at separation, and it in turn carries on to significant growth of secondary ablation of a dust.

The flow of dusty gas passing through blade impellers, it is possible to schematize a plane flow of incompressible liquid which approaches to the mesh at an angle α with in a speed ϑ and at turn about crimps of blades loses dust corpuscles, rebounding from blades. Purified gas is torn off from crimps of blades, forming in slotted channels between them fly off zones with approximately fixed pressure. Boundary lines of fly off zones can be observed as a line of rupture of tangential speeds in an ideal (nonviscous) fluid. Definition of fields of speeds of gas at transiting through blade impellers inconveniently enough. The calculation problem can be simplified, if at research of process of separation of a dust to observe the approximate model of the gas flow considering basic regularity installed at more exact calculation of fields of speeds of dust-laden gas medium.

Thus, if design data of blades of an impeller are set, it is possible to count a gas velocity distribution at flow by its dust-laden gas flow. Knowing speeds of gas in various points of a gas stream, it is possible to determine forces of an aerodynamic resistance on which mechanical trajectories of hard corpuscles in slotted channels depend.

4.4   RESEARCH OF AGENCY OF BLADES OF THE IMPELLER OF THE N HYDRODYNAMIC CIRCUMSTANCES IN THE APPARATUS

Judging by expositions in the literature, the bending of blades pursues the aim to excite centrifugal forces and to use them for separation of corpuscles within the curved channel—an impeller. The great value concerning a dust separation is assigned also to curtains of drops and the splashes, doubly crossing an airflow path.

Considering, as further it will be shown, also very low water resistance of sinusoidal blades of an impeller with an intensive twisting of air flows in their each element, it is possible to draw a leading-out that similar blades have good prospects for the solution warmly—and mass transfer problems in system gas-fluid.

The physical picture of interacting of air and fluid in blade impellers is introduced on Fig. 4.3.

FIGURE 4.3   Gas and irrigating liquid interacting.

Computer simulation has been applied to study of hydrodynamic performances in the apparatus in program ANSYS-14 CFX. For calculation the finite element method was used. Flow of gas without a discrete phase in view of that power and heat effect of corpuscles on gas phase flow negligibly is not enough has been simulated only. This conjecture is true in that event when the mass fraction of corpuscles in a two-phase flow does not exceed 30% [2]. The Numerical analysis of flow of gas in the shock-vortical apparatus is reduced to the solution of system mean on Reynolds of the equations of the Navier-Stokes. For gas dynamic short circuit, the equations of the Navier-Stokes it was used standard k–e turbulence model. Besides turbulence and heat exchange model, it was used VOF (Volume of Fluid) for simulation with a water and air flow free surface.

Boundary conditions have been chosen the following: in the capacity of gas has been used – air, fluids on which air – water goes, temperature on an entry – 300K, pressure of gas on an exit hardly above atmospheric – 101,425 passes, intensity of turbulence on an entry and an apparatus exit – 5%.

The first investigation phase was definition of a role of blades of an impeller. For this purpose comparative researches on a dust clearing efficiency in one apparatus have been conducted at two constructions:

Ordinary rotoklon (a Fig. 4.4); a rotoklon with the blade impellers offered in the present activity (a Fig. 4.5).

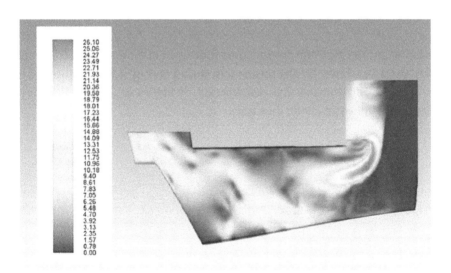

FIGURE 4.4   Gas and irrigating liquid interacting.

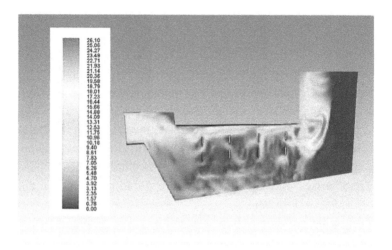

FIGURE 4.5   Gas and irrigating liquid interacting.

If for an ordinary rotoklon efficiency of trapping of a dust was small and strongly depended on level of the fluid which are filled up in the apparatus, in the presence of impeller blades efficiency. Trapping in all velocity band of air was high and depended on fluid level in the apparatus (*see* Figs. 4.6 and 4.7) a little. By the results introduced on Fig. 4.8, it is possible to draw a leading-out that the construction of blades of an impeller plays large role in inertia-shock dust traps.

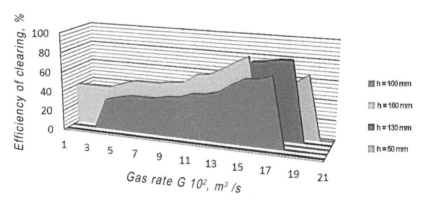

FIGURE 4.6   Relation of efficiency of clearing of gas to airflows and an irrigating liquid without blades.

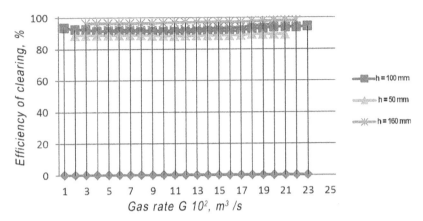

FIGURE 4.7   Relation of efficiency of clearing of gas to airflows and an irrigating liquid with sinusoidal blades.

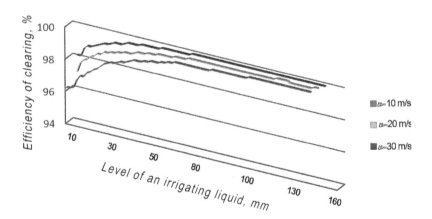

FIGURE 4.8   Relation of efficiency of clearing of gas to level of a filling up of a fluid at various speed of gas.

4.5   THE ESTIMATION OF THE MAXIMUM EXTENT OF RECIRCULATION OF THE IRRIGATING LIQUID

4.5.1   STATE-OF-THE-ART OF A PROBLEM

It is possible to admit that for certain three-phase systems gas—a liquid—a solid boundary dust load can appear too large that directly will affect too large extent of circulation which is difficult for realizing in the conditions of commercial operation of wet-type collectors.

The problem formulation leans on following rules. In the conditions of full circulation of a fluid, at fixed geometrical sizes of a dust trap it is possible to supply a persistence of operation parameters is a relative speed of driving of a fluid and an aerosol, dust load in gas, surface tension of a fluid or an angle of wetting of a dust. Dust load growing in time in a fluid carries on to unique essential change—to increase in its viscosity. After excess of certain density suspension loses properties of a Newtonian fluid. Dust trap working conditions at full circulation of a fluid are approximate to what can be gained in the batch action apparatus when at maintenance in a dust separation system fresh water is not inducted. Collected in the apparatus, the dust detained by a fluid, compensates volumetric losses of the fluid necessary on moistening of passing gas and its matching ablation. In the literature there are no activities, theoretically and experimentally presenting agency of viscosity and agency of rheology properties of suspension on a dust clearing efficiency. As it seems to us, a motive is that fact that in the capacity of operating fluid water is normally used, and dust traps work, in the core, at fixed temperatures. Simultaneously, at use of fractional circulation certain level of dust loads in a fluid is supplied. In turn, accessible relations in the literature indicate negligible growth of viscosity of suspension even at raise of its density for some percent.

The reasoning's justifying a capability of agency of viscosity of suspension on a dust clearing efficiency, it is possible to attribute as on the analysis of the basic gears influencing sedimentation of corpuscles on an interphase surface, and on conditions of formation of this developed surface of a fluid. Transition of corpuscles of a dust from gas in a fluid occurs, primarily, as a result of the inertia affecting, effect of "attachment" and diffusion. Depending on type of the wet-type collector of a corpuscle of a dust deposit on a surface of a fluid which can be realized in the form of drops, moving in a flow of an aerosol, the films of a fluid generated in the apparatus, a surface of the gas vials organized in the conditions of a splashing and moistened surfaces of walls of the apparatus.

One of trends of development of wet-type collectors is creation of apparatuses of intensive operation with high carrying capacity on a gas phase that is linked with expedient lowering of gabarits of installations. In these conditions owing to high relative speed of driving of the liquid and gas phases, solving agency on effect of a dust separation is rendered by gears: the inertia and immediate acquisition of corpuscles. Such process is realized in the inertia dust traps to which the examined apparatus concerns.

In the monography [3] agency of various gears on efficiency of sedimentation of corpuscles of a dust on fluid surfaces is widely presented. The exposition of gears and their agency on a dust clearing efficiency can be discovered practically in all monographies, for example [4], concerning a problem wet dust separations gases. In the literature of less attention it is given questions of shaping of surfaces of fluids and their agency on a dust clearing efficiency. Observing the gear of the inertia act irrespective of a surface of the fluid capturing a dust, in the core it is considered that for hydrophilic types of a dust collision of a part of a solid with a fluid surface to its equivalently immediate sorption by a fluid, and then immediate release and a refacing of a fluid for following collisions. In case of a dust badly moistened, the time necessary for sorption of a corpuscle by a fluid, can be longer, than a time after which the corpuscle will approach to its surface. Obviously, it is at the bottom of capability lowering retardations a dust a fluid because of a recoil of the corpuscle going to a surface, from corpuscles which are on it. It is possible to consider this effect real as in the conditions of a wet dust separation with a surface of each fluid element hits has more dust, than it would be enough for monolayer shaping. Speed of sorption of corpuscles of a dust can be a limiting stage of a dust separation. Speed of sorption of a corpuscle influences not only its energy necessary for overcoming of a surface tension force, but also and speed of its driving in the liquid medium, depending on its viscosity and rheology properties. Dust clearing efficiency Kabsch [5] links with speed of hobby of a dust a fluid, having presented it as mass m_s, penetrating in unit of time through unit of a surface And in depth of a fluid as a result of a collision of grains of a dust with this surface

$$r = \frac{m_s}{A \cdot t}$$

Speed of linkage of a dust a fluid depends from fiziko—chemical properties of a dust and its ability to wetting, physical and chemical properties of gas and operating fluid, and also density of an aerosol. Wishing to confirm a pushed hypothesis, Kabsch [5] conducted the researches concerning agency of density speed of linkage of a dust by a fluid. The dust load increase in gas called some increase in speed of linkage, however to a lesser degree, than it follows from a linear dependence.

The cores for engineering of a wet dust separation of model Semrau, Barth'a and Calvert'a do not consider agency of viscosity of suspension on effect of a dust separation. Speed of driving can characterize coefficient of resistance to corpuscle driving in a fluid, so, and a coefficient of impact of viscosity of a fluid. Possibility of effect of viscosity of a liquid on efficiency of capture of corpuscles of a dust a drop by simultaneous act of three mechanisms: the inertial, "capture" the semiempirical equation Slinna [9] considering the relation of viscosity of a liquid to viscosity of gas also presents.

In general it is considered that there is a certain size of a drop at which optimum conditions of sedimentation of corpuscles of a certain size are attained, and efficiency of subsidence of corpuscles of a dust on a drop is sweepingly reduced with decrease of a size of these corpuscles.

Jarzkbski and Giowiak [7], analyzing activity of a shock-inertial dust trap have installed that in the course of a dust separation the defining role is played by the phenomenon of the inertia collision of a dust with water drops. Efficiency of selection of corpuscles of a dust is reduced together with growth of sizes of the drops oscillated in the settling space, in case of a generating of drops a compressed air, their magnitude is determined by equation Nukijama and Tanasawa [8] from which follows that drops to those more than above value of viscosity of a liquid phase. Therefore viscosity growth can call dust clearing efficiency reduction.

Summarizing, it is possible to assert that in the literature practically there are no data on agency of viscosity of a trapping fluid on dust separation process. Therefore one of the purposes of our activity was detection of agency of viscosity of a fluid on a dust clearing efficiency.

4.5.2   THE PURPOSE, CONJECTURES AND AREA OF RESEARCHES

The conducted researches had a main objective acknowledging of a hypothesis on existence of such boundary density of suspension at which excess the overall performance of the dust removal apparatus is reduced.

The concept is developed and the installation which is giving the chance to implementation of scheduled researches is mounted. Installation had systems of measurement of the common and fractional efficiency and standard systems for measurement of volume-flows of passing gas and hydraulic resistance. The device of an exact proportioning of a dust, and also the air classifier separating coarse fractions a dust on an entry in installation is mounted. The conducted gauging of measuring systems has supplied respective recurrence of the gained results.

4.5.3   THE LABORATORY-SCALE PLANT AND A TECHNIQUE OF REALIZATION OF EXPERIMENT

Laboratory-scale plant basic element is the dust trap of shock-inertial act—a rotoklon c adjustable lobes (Fig. 4.9). An aerosol gained by dust injection in the pipeline by means of the batcher (Fig. 4.9). Application of the batcher with varying productivity has given the chance to gain the set dust load on an entry in the apparatus.

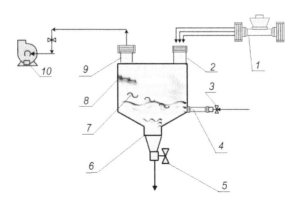

FIGURE 4.9   The Laboratory-scale plant.

Have been examined a dust, discriminated with the wettability (Talcum TMP) [9], median diameter is equal $\delta_{50} = 25$ microns, white black about δ_{50} = 15 microns solubility in water of 10^{-3}% on mass (25°C) and a chalk dust. Dust flow passed blade impellers 7 in an apparatus working area, whence through the entrainment separator 8 cleared, was inferred outside. Gas was carried by means of the vacuum pump 10, and its charge measured by means of a diaphragm 1. A Gas rate, passing through installation, controlled, changing quantity of air sucked in the pipeline before installation. Gas differential heads were measured by the pressure gauge. Dust load in entering and getting out gases measured by means of the analogous systems introduced earlier. The composition of each system includes group of three sondes mounted on vertical sections of pipelines, on distance about 10 diameters from the proximal element changing the charge. The taken test of gas went on the measuring filter on which all dust contained in test separated. For this purpose used a filter paper of type AFA, including it as the filter absolute. In the accepted solution have applied system of three measuring sondes which have been had in pipelines so that in minimum extent to change a regime of transiting of gas and to take from quantity of a dust necessary for the analysis. The angle between directions of deduction of sondes made 120°, and their ends placed on such radiuses that surfaces of rings from which through a sonde gas was sucked in, were in one plane. It has allowed to divide out a time of selection of test and gave average dust loads in gas pipeline cross-section.

For definition of structurally mechanical properties of suspension viscosity gauge RV-8 has been used. The viscosity gauge consists of the inner gyrating cylinder (rotor) ($r = 1.6$ sm) and the external motionless cylinder (stator) ($r = 1.9$ sm), having among themselves a positive allowance of the ring form with a size 0.3 see the Rotor is resulted in twirl by means of the system consisting of the shaft, a cone ($K = 2.23$ sm), filaments, blocks and a cargo. To twirl terminations apply a brake. The gyrating cylinder has on a division surface on which control depth of its plunging in suspension.

The gained suspension in number of 30 sm³ (in this case the rotor diving depth in slurry makes 7 sm) fill up in carefully washed up and dry external glass which put in into a slot of a cover and harden its turn from left to right. After that again remove the loaded cylinder that on a scale of the inner cylinder precisely to determine depth of its plunging in slurry.

Again fix a glass and on both cups lay a minimum equal cargo (on 1), fix the spigot of a cone by means of a brake and reel a filament, twirling a cone clockwise. Monitor that convolutions laid down whenever possible in parallel each other.

Install an arrow near to any division into the limb and, having hauled down a brake, result the inner cylinder in twirl, fixing a time during which the cylinder will make 4–6 turnovers. After the termination of measurements fix a brake and reel a filament. Measurement at each load conduct not less than three times. Experience repeats at gradual increase in a cargo on 2 r until it is possible to fix a time of an integer of turnovers precisely enough. After the termination of measurements remove a glass, delete from it slimes, wash out water, from a rotor slimes delete a wet rag then both cylinders are dry wiped and leave a gear in the collected aspect.

After averaging of the gained data and angular rate calculation the graph of relation of twirl rate from the applied load is under construction

Viscosity is determined by formula:

$$\eta = \frac{(R_2^2 - R_1^2)Gt}{8\pi^2 L R_1^2 R_2^2 L} \tag{1}$$

$$\text{or } \eta = \frac{kGt}{L} \tag{2}$$

4.5.4 RESULTS OF RESEARCHES

For each dust used in researches the relation of general efficiency of a dust separation to density of suspension and the generalizing graph of relation of fractional efficiency to a corpuscle size is introduced. Other graph grows out of addition fractional dust clearing efficiency for various, introduced on a drawing, densities of suspension. In each event the first measurement of fractional efficiency is executed in the beginning of the first measuring series, at almost pure water in a dust trap. In activity curve flows of water suspensions studied a dust are introduced. Curve flows presented sedate model Ostwald-de Waele in the field of applied speeds of shift. From character of curves follows that for chalk suspension, for density of corpuscles of a solid less than 12% mass suspension behaves as a Newtonian fluid, with growth of density suspension behaves as non-Newtonian—a pseudo-plastic fluid. Suspension talcum with density less than 40% mass behaves as a Newtonian fluid. For densities above 45% mass suspension starts to show features of non-Newtonian fluid accurately. Water suspensions of carbon black with density less than 7% mass behave as Newtonian fluids. For density about 12% for speed of shift less than 400 seconds show weak properties of non-Newtonian fluid. Analyzing the gained results of researches of general efficiency of a dust separation, it is necessary to underscore that in a starting phase of activity of a rotoklon at negligible density of suspension for all used in researches a dust separation high performances, components from 93,2% for carbon black to 99,8% for a talc dust are gained. Difference of general efficiency of trapping of various types of a dust originates because of their various particle size distributions on an entry in the apparatus, and also because of the various form of corpuscles, their dynamic wettability and density. The gained high values of general efficiency of a dust separation testify to correct selection of constructional and operation parameters of the studied apparatus and indicate its suitability for use in engineering of a wet dust separation.

The important summary of the conducted researches was definition of boundary densities of suspension various a dust after which excess general efficiency of a dust separation is reduced. Value of magnitude of boundary density, as it is known, is necessary for definition of the maximum extent of recirculation of an irrigating liquid. As appears from introduced in drawings 8.11–8.14 graphs, the relation of general efficiency of a dust separation to density of suspension, accordingly, for a dust of talc, a chalk and white black is available a capability of definition of such densities.

Boundary densities for talcum—36%, white black—7%, a chalk—of 18% answer, in the core, to densities at which suspensions lose properties of a Newtonian fluid.

The conducted researches give grounds to draw leading-outs that in installations of shock-inertial type where the inertia gear is the core at selection of corpuscles of a dust from gas, general efficiency of a dust separation essentially drops when density of suspension answers such density at which it loses properties of a Newtonian fluid. As appears from introduced in Figs. 4.10–4.13 relations, together with growth of density of suspension above a boundary value, general efficiency of a dust separation is reduced, and the basic contribution to this phenomenon shallow corpuscles with a size less bring in than 5 microns. To comment on the relations introduced in drawings, than 5 microns operated with criterion of lowering of a dust clearing efficiency of corpuscles sizes less at the further increase in density of suspension at 10% above the boundary. Having taken it in attention, it is possible to notice that in case of selection of a dust of talc growth of density of suspension from 36% to 45% calls reduction of general efficiency of a dust separation from 98% to 90% at simultaneous lowering of fractional efficiency of selection of corpuscles, smaller 5 microns from $\eta = 93\%$ to $\eta = 65\%$.

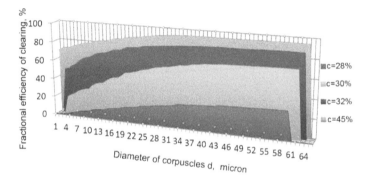

FIGURE 4.10   Relation of fractional efficiency to diameter of corpuscles TMP and their densities in a fluid.

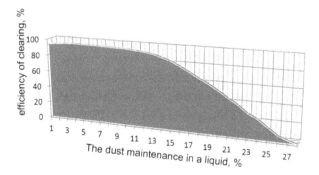

FIGURE 4.11   Relation of general efficiency of density in a fluid of corpuscles TMP.

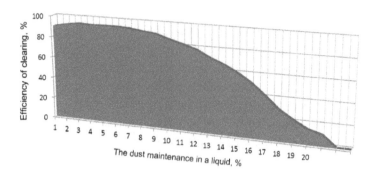

FIGURE 4.12   Relation of fractional efficiency to diameter of corpuscles of white black and their density in a fluid.

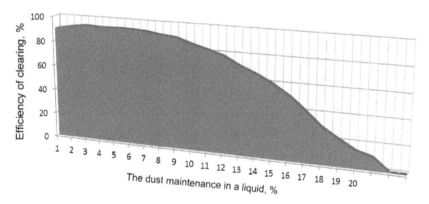

FIGURE 4.13   Relation of general efficiency of density in a fluid of corpuscles of white black.

Analogously for white black: growth of density from 7% to 20% calls fall of fractional efficiency from $\eta = 65\%$ to $\eta = 20\%$, for a chalk: growth of density from 18% to 30% calls its lowering from $\eta = 80\%$ to $\eta = 50\%$. Most considerably lowering of a fractional dust clearing efficiency can be noted for difficultly moistened dust—white black (about 50%).

Thus, on the basis of the analysis set forth above it is possible to assert that lowering of general efficiency of a dust separation at excess of boundary density of suspension is linked about all by diminished ability of system to detain shallow corpuscles. Especially it touches badly moistened corpuscles. It coincides with a hypothesis about renewal of an interphase surface. Renewal of an interphase surface can be linked also with difficulties of driving of the settled corpuscles of a dust deep into fluids, that is, with viscosity of medium.

The analysis of general efficiency of a dust separation, and, especially, a dust talcum and white black indicates that till the moment of reaching of boundary density efficiency is kept on a fixed level. In these boundary lines, simultaneously with growth of density of suspension the fluid dynamic viscosity, only this growth grows is negligible—for talc, for example, to 2.7×10^{-3} Pa × s. At such small increase in viscosity of suspension the estimation of its agency on a dust clearing efficiency is impossible. Thus, it is possible to assert, what not growth of viscosity of suspension (from 1 to 2.7×10^{-3} Pa × s), and change of its rheology properties influences dust clearing efficiency lowering.

The method of definition of boundary extent of circulation of a fluid in shock-inertial apparatuses is grounded on laboratory definition of density of suspension above which it loses properties of a Newtonian fluid. This density will answer density of operating fluid, which cannot be exceeded if it is required to supply a dust clearing efficiency constant.

In the conditions of spent researches, that is, fixed density of an aerosol on an entry in the apparatus, and at the conjecture that leakage of water in the apparatus because of moistening of passing air and, accordingly, ablation in the form of drops, is compensated by a collected dust volume, the water discharge parameter is determined directly from recommended time of duration of a cycle and differs for various types of a dust. Counted its maximum magnitude is in the interval 0.02–0.05 l/m³, that is, is close to the magnitudes quoted in the literature.

For batch action dust traps this density determines directly a cycle of their activity. In case of dust traps of continuous act with fluid circulation, the maximum extent of recirculation supplying maintenance of a fixed level of a dust clearing efficiency, it is possible to count as a ratio

$$r = \frac{Q_{cir}}{Q_{ir}} \tag{3}$$

where, Q_{cir} is the charge recirculation a liquid; Q_{ir} is the charge of a liquid on irrigating.

Assuming that all dust is almost completely trapped on a fluid surface, it is possible to record a balance equation of mass of a dust as:

$$G \cdot (c_{on} - c_{in}) = L \cdot c_{on} \tag{4}$$

where $(c_{on} - c_{in}) = c_r$ is the limiting dust load, g/m³.

Then taking into account of Eq. (4) for calculation of extent of recirculation it is possible to introduce as:

$$r = \frac{G \cdot c_{on}}{L \cdot c_r} \tag{5}$$

4.6   CONCLUSIONS

1. Experimentally it is installed that aerodynamic profiling of blades of an impeller on a sinusoid allows to reduce a device water resistance considerably. Thus there is an increase in efficiency of clearing of gas thanks to flow of an input section of a profile of blades with large fixed speed and to increase in skips from different parts of blades.
2. Excess of boundary density of suspension at which it loses properties of a Newtonian fluid, calls dust clearing efficiency lowering.
3. At known boundary density c_r it is possible to determine boundary extent of recirculation of the irrigating liquid, supplying stable dust clearing efficiency
4. Magnitude of boundary density depends on physical and chemical properties of system a fluid—a solid and varies over a wide range, from null to several tens percent. This magnitude can be determined now only laboratory methods.
5. Lowering of an overall performance of the apparatus at excess of boundary density is linked, first of all, with reduction of fractional efficiency of trapping of shallow corpuscles dimensioned less than 5 microns.
6. Selection constructional and the operating conditions, securing a dust separation high performance at small factor of water

consumption, allow recommending such dust traps for implementation in the industry.

The leading-outs formulated earlier are actual for intensive operation wet-type collectors in which the basic gear of selection of corpuscles is the gear of the inertia dust separation.

KEYWORDS

- calculation of operating conditions
- clearing of gas
- irrigating liquid
- recirculation
- rotoklon

REFERENCES

1. Patent 2317845 RF, IPC, cl. B01 D47/06 *Rotoklon a controlled sinusoidal blades*, R. R. Usmanova, ZhernakovV.S, Panov, A. K.-Publ. 27.02.2008. Bull. № 6.
2. Uzhov, V. N., Valdberg, A. *Treatment of industrial gases from dust*. V. N. Uzhov, A. U.Valdberg, Chemistry, Moscow, 1981, 280 p.
3. Straus, V. *Industrial gas cleaning /*. V. Straus, Chemistry, Moscow, 1981, 616 p.
4. Shwidkiy, V. S. *Purification of gases. Handbook*. V. S. Shwidkiy Thermal power, Moscow, 2002, 640 p.
5. Kabsch-Korbutowicz, M., Majewska-Nowak, K., *Removal of atrazine from water by coagulation and adsorption*. Environ Protect Eng. 29 (3–4) (2003), 15–24.
6. Kitano, T., and Slinna, T. "*An empirical equation of the relative viscosity of polymer melts filled with various inorganic fillers*". (1981), Rheologica Acta 20 (2): 207.
7. Jarzkbski, L. Giowiak, An. Ochrvmy Ibrod. itr, II (1977).
8. Nukijama, S., Tanasawa, Y., Trans. Soc. Mech. Eng. Japan. 1999. v. 5.
9. GOST 21235-75 *Talcum. Specifications*.

CHAPTER 5

THE EFFECT OF THE MODIFICATION OF SILICA-GELATIN HYBRID SYSTEMS ON THE PROPERTIES OF SOME PAPER PRODUCTS

PRZEMYSŁAW PIETRAS, ZENON FOLTYNOWICZ, HIERONIM MACIEJEWSKI, and RYSZARD FIEDOROW

CONTENTS

5.1. INTRODUCTION

Silica-gelatin hybrid systems obtained by sol-gel method using two silica precursors: tetraethoxysilane (TEOS) and ethyl silicate (U740) as well as two kinds of gelatin (FGG and RGG) were studied. The study was aimed at evaluating the hybrid systems as impregnates for paper and cardboard. For this reason two kinds of paper and two kinds of cardboard were subjected to modification with the hybrid systems and the evaluation was performed of the change in such parameters as grammage, durability on elongation, durability on rupture, absorptiveness of water and burning time. Results of the study have shown a positive effect of the impregnates on functional parameters of paper products nonmodified by the manufacturers during the production process.

Since the beginning of papermaking in Western Europe, gelatin was used to reduce the sorption of water (paper sizing) by papers in order to improve the buffer effect and feathering of the inks [1]. Nowadays, papers are sized with synthetic sizes such as alkyl ketene dimers (AKDs) and alkenyl succinic anhydrides (ASAs), which were developed for the paper industry in 1953 and 1974, respectively [2], but gelatin sizing continues to be used for artist quality papers.

Hybrid materials consisting of biopolymer and silica raise hopes for obtaining materials of high compatibility to tissues and interesting properties. Such hybrids find application as biocompatible materials, bone substituents, as well as immobilizers of enzymes, catalysts and sensors, preceramic materials and many others [3, 5, 6]. A very convenient and effective method for preparing such systems is sol-gel technique. Biocomposites obtained by the mentioned method have exceptional properties such as high plasticity, low modulus of elasticity and high strength [3–7].

Gelatin is a product of thermal denaturation of collagen present in bones and animal skins. It is formed as a result of temperature action on three-chain polypeptide helix of collagen, which undergoes unwinding to give gelatin balls. It is not a homogeneous product and its composition depends on collagen origin. Molecular weight of gelatin ranges between 80,000 and 200,000. Gelatin is a fully biodegradable material [8–11]. It dissolves in water after heating to about 40°C. Gelatin has its greatest application in food, pharmaceutical and photographic industries. Its potential applications include bone reconstructions, biocatalysis, drug distribution control in living organisms, etc. [3, 12].

In the case of hybrid materials, of great importance is the formation of durable bonds between biopolymer and inorganic gel. The fundamental role in the formation of such bonds is played by silane coupling agents such as aminosilanes, glycidoxysilanes, etc. [13]. For gelatin-containing systems the most frequently selected coupling agent is glycidoxypropyltrimethoxysilane. Ren et al. [14, 15] described synthesis of porous siloxane derivatives of gelatin and 3-glycidoxypropyltrimethoxysilane (GPTSM), prepared by sol-gel technique in the presence of HCl. In another paper Ren et al. [16] described synthesis of gelatin—siloxane hybrids with gelatin bound terminally to siloxane chains, using GPTSM as a coupling agent. Smitha et al. [17] presented a method for immobilization of gelatin molecules in mesoporous silica by carrying out tetraethoxysilane (TEOS) hydrolysis under conditions of controlled pH. Pietras et al. [18] described preparation of silica-gelatin hybrid materials by sol-gel method using ethyl silicate (U740) as a silica precursor and GPTMS as a coupling agent.

In this chapter, syntheses of silica-gelatin hybrid materials is aimed at obtaining possibly the highest degree of compatibility of the systems prepared by the integration of two types of polymers—organic biopolymer (gelatin) and inorganic polymer (silica gel) using 3-glycidoxypropyltrimethoxysilane (GPTSM) as a coupling agent. The obtained hybrid materials were applied as modifiers of paper products. The modified paper products were characterized from the point of view of the influence of employed impregnates on such functional parameters as basis weight change, durability on elongation, durability on rupture, absorptiveness of water, burning time.

5.2 MATERIALS AND METHODS

5.2.1 MATERIALS

Gelatin employed in the study originated from porcine skin (type A, isoelectric point (pI) = 8, 300 Bloom, dry matter 90%) was purchased from Aldrich (denoted as RGG, research grade gelatin) and edible porcine gelatin was purchased from Kandex, Ltd. (denoted as FGG, food grade gelatin). Ethyl silicate—U740 (40% SiO_2), tetraethoxysilane (TEOS) and 3-glycidoxypropyltrimethoxysilane (GPTMS) were obtained from Unisil Ltd. (Tarnów, Poland) and glacial acetic acid from Aldrich.

The modification was performed on white paper Lux 80 g/m² for black and white printing and copying, manufactured by International Paper POL, Poland (**PB**), as well as cardboard for painting and building purposes, manufactured by Blue Dolphin Tapes, Poland (**TM**). To compare properties of the impregnated paper materials with those commercially available, we have also subjected to impregnation a packing paper manufactured by Hoomark, Ltd., Jędrzychowice, Poland (**PP**) and a cardboard A1–170 g/m² manufactured by Kreska, Bydgoszcz, Poland (**TZ**).

5.2.2 SILICA-GELATIN HYBRID MATERIALS

The starting sol was obtained by hydrolysis and condensation of an appropriate amount of ethyl silicate (U740, d=1.05 g/cm³, 40% SiO_2) or tetraethoxysilane (d=0.933 g/cm³, 28.85% SiO_2) and GPTMS in water in the presence of acetic acid as pH control agent. In the typical experiment 166 g of TEOS or 120 g of U740, 61 g of GPTMS and 60 g of acetic acid were used. The amount of silica formed as a result of hydrolysis and condensation of both precursors was constant and taken into account each time when proportions of ethyl silicate or tetraethoxysilane to 3-glycidoxypropyltrimethoxysilane were chosen. The mixture was vigorously stirred and heated at 85°C for 6h to yield functionalized silica gel. At the next stage, gelatin was added to the obtained sol and the mixture was heated at 60°C for 2h with vigorous stirring. Codes and compositions of samples obtained are shown in Table 5.1. Then this gel was applied as the impregnate for paper products.

TABLE 5.1   Percentage of gelatin* and codes of silica-gelatin hybrid materials based on TEOS and U740.

Gelatin added [%]	TEOS-based sample coding	U740-based sample coding
0	T0	U0
50	T1	U1
100	T2	U2
150	T3	U3

*Calculated with reference to theoretical amount of silica formed from TEOS or U740.

5.2.3   MODIFICATION OF PAPER PRODUCTS

A 350 mL of the impregnate (ca. ¼ of obtained silica-gelatin hybrid system) was placed into a photographic tray. Six sheets of paper of A4 size or cardboard were modified in a given portion of the impregnate. The impregnation was carried out by deep coating method for 5 minutes (FGG5 and RGG5) or for 15 minutes (FGGS15 and RGG15). Then the impregnated sample was hung over the tray to enable the excess impregnate to flow away. This was followed by drying of the sample in an air recirculation oven at 60°C for 30 minutes. The preparations obtained were designated as FGG5, FGG15, RGG5 and RGG15, respectively.

5.2.4   CHARACTERIZATION OF THE SYSTEMS

The obtained materials were characterized from the point of view of changes in functional parameters such as basis weight change, durability on elongation, durability on rupture, absorptiveness of water, and burning time.

Changes in grammage were measured as follows: samples of 1000 mm² area were cut out of modified paper with the accuracy of 0.5 mm². Each sample was weighed on a balance with the accuracy of three significant digits. Measurements were performed for 10 samples of each modified paper and results were averaged. Durability on elongation was determined using tensile testing machine VEB TIW Rauenetein ZT; working length was

100 mm, sample width 15 mm, elongation speed 20 mm/min. Durability on rupture of cardboards weighing over 250 g/cm^2 was measured with the use of an apparatus made by Lorentzen & Wettre Co. at initial pressure of 0.5 MPa. Absorptiveness of water was measured by the Cobb method in compliance with the standard ASTM D3285-93 (2005) by determining the Cobb60 index; sample surface area was 100 cm^2. Burning time was measured under controlled conditions in a closed chamber equipped with a burner for combustion of samples; height of the burner flame was 1.5–2 cm.

5.3   RESULTS AND DISCUSSION

5.3.1   *CHARACTERIZATION OF MODIFIED PAPER PRODUCTS*

Within the study two series were synthesized of silica-gelatin hybrid systems based on TEOS and U740 with increasing gelatin (FGG or RGG) content. The prepared silica-gelatin hybrid systems were used for the impregnation of paper products PB, TM, PP and TZ in order to evaluate the effect of silica precursors applied in the study, as well as that of origin and concentration of gelatin on selected qualitative and functional parameters of the paper products studied.

All samples of paper products maintained their initial shape and size after impregnation and drying and no curling up occurred. However, an increase in the rigidity of samples was observed. The use of the impregnate had no influence whatsoever on the possibility of writing with a ball pen, pencil or felt-tip pen on the samples.

Organoleptic evaluation with naked eye did not show color changes in the investigated paper samples, except for TM in the case of which color darkening was observed. No separation of impregnate from paper was noticed during cutting of modified paper samples into smaller pieces required for different analyzes. The samples have retained their shape after cutting—they did not curl up.

5.3.2   BASIS WEIGHT CHANGE

Changes in the grammage of PB samples modified with hybrid systems based on TEOS as well as FGG and RGG were shown in Fig. 5.1. One can notice that grammage of a given modified paper increases compared to that of nonmodified (**NM**) paper. Analogous situation was observed for all paper samples impregnated with all hybrid systems studied. Initial grammages of investigated papers were 80 g/m^2 for PB, 289 g/m^2 for TM, 51 g/m^2 for PP and 170 g/m^2 for TZ. The highest values of impregnated paper grammage were observed for the last samples of a given series modified with systems based on TEOS and U740 with the greatest FGG or RGG gelatin content. The grammage of PB increased to reach the highest value of 110 g/m^2 for the sample T3 FGG5. In the case of TM, analogously to the above mentioned paper samples, the grammage raises with increasing gelatin content to the maximum value of 363 g/m^2 in the case of the last sample of FGG15 series based on TEOS. The raise in grammages with increasing gelatin content in impregnates was also observed for all control paper samples. In the case of PP, the grammage increases to 71 g/m^2, while for TZ to 194 g/m^2. The increase in grammages of all samples of all series of papers modified with hybrid materials compared to starting papers testifies for the penetration of impregnates used into papers investigated This fact is additionally proved by the appearance and behavior of samples during their preparation and analyzes (impregnate was not coming off paper during cutting out and no splinters were formed while testing strength of samples).

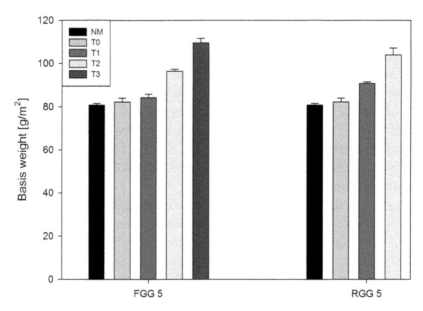

FIGURE 5.1   Grammage of nonmodified (NM) PB and TEOS-modified (T0-T3) PB.

5.3.3 DURABILITY ON ELONGATION

Durability on elongation is the maximum tensile force acting on width unit of a sample at which the sample remains unbroken. The measurement of the above parameter consisted in placing a sample in holders of tensile testing machine and stretching the sample with a constant speed until the latter underwent breaking. The value of tensile force at the moment of breaking was recorded. Averaged results of measurements of breaking resistance of PB impregnated with hybrid materials based on TEOS and FGG as well as RGG were shown in Fig. 5.2. In the case of PB and TM an improvement in this parameter was observed, whereas in that of PP and TZ a slight reduction in durability on elongation was noticed. Values of the above parameter were 3.8 kN/m, 5.4 kN/m, 6.2 kN/m and 9.4 kN/m for unmodified PB, TM, PP, and TZ, respectively. The durability on elongation of impregnated PB increased with a rise in gelatin content for all series of both hybrid systems. The highest values were observed for the last samples of a given series and were 6.0 kN/m, 6.2 kN/m, 5.3 kN/m and 5.1

kN/m for FGG5 and RGG5 series for hybrid systems based on TEOS and U740, respectively. Similarly as it was in the case of PB, also in that of TM an increase was observed in the durability on elongation. The mentioned parameter has the highest value for T3 sample of FGG5 series and equals to 9.1 kN/m for hybrid systems based on TEOS and for U3 sample of FGG5 series being equal to 9.4 kN/m for hybrid systems based on U740. In contradistinction to previous paper products, in the case of PP and TZ a small decrease was observed in the durability on elongation. In the case of PP, the highest reduction in this parameter occurs for T1 sample of RGG15 series and equals to 4.5 kN/m, while for other samples of all series it ranges between 5.2–6.0 kN/m. Results obtained for TZ show that the highest decrease in the durability on elongation takes place for T2 sample from FGG5 series and is equal to 7.8 kN/m, while for most of samples from other series it is in the range of 8–9 kN/m.

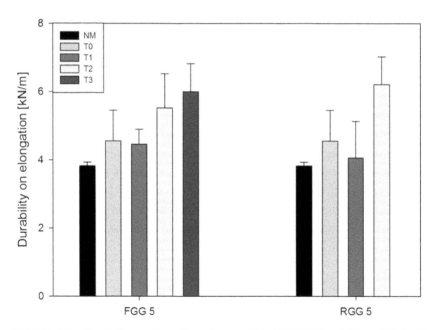

FIGURE 5.2 Durability on elongation of nonmodified (NM) PB and PB modified with TEOS (T0-T3).

5.3.4   DURABILITY ON RUPTURE

Burst is the strength of a single sheet of paper to withstand homogeneous pressure acting perpendicularly to the surface of the sheet. It is the basic parameter for evaluation of strength of packing paper and corrugations of corrugate board. The measurement consists in tightening a paper sample between two rings of flexible membrane, which undergoes bulging under the effect of pressure of air delivered until the sample breaks. Bursting strength of a sample corresponds to maximum value of pressure that resulted in the sample breakage. The measurement was carried out for ten TM-modified samples and results were averaged. Fig. 5.3 shows averaged results for TM modified with hybrid systems based on U740 and FGG as well as RGG. It is clearly seen in the figure that bursting strength of impregnated TM is considerably greater than that of nonimpregnated TM for which the burst was 252 kPA. In the case of cardboard modified with hybrid systems based on U740 an increase in bursting strength is noticeable for both FGG series with the rise in gelatin content. The values change from 350 kPa to 531 kPa and 359 kPa to 482 kPa for FGG5 and FGG15 series, respectively. For both RGG series a decrease in bursting pressure was observed in the case of gelatin-containing samples compared to samples impregnated with pure silica sol, however, the decrease was small (from 539.0 kPa to 514.5 kPa and from 414.5 kPa do 393.0 kPa for RGG 5 and RGG15 series, respectively). In the case of samples modified with TEOS-based hybrid systems the bursting strength increased as much as to 555.5 kPa for samples from FGG5 and RGG5 series and to 548.0 kPa for those from RGG15 series. Summing up, one can conclude that the impregnation of TM with the studied hybrid systems resulted in the increase in bursting strength, which in some cases was over 100%.

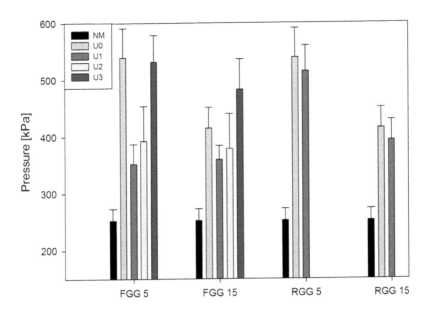

FIGURE 5.3 Durability on rupture of nonmodified (NM) TM and TM modified with U740 (U0-U3).

5.3.5 ABSORPTIVENESS OF WATER

Absorptiveness of water is the mass of water absorbed by 1 m² of paper at a specified time and specified conditions. The measurement consists in determining the mass of water absorbed by a paper sample of 100 cm² area in a specified time under the water level of 1 cm. Measurements were performed for 10 samples of each modified paper and results were averaged. The change in absorptiveness of water for PB samples modified with hybrid systems based on TEOS and FGG as well as RGG was presented in Fig. 5.4. In the case of PB and TM an improvement in this parameter was observed, whereas in that of PP and TZ a considerable increase occurred in water absorptiveness. The relevant values were 95 g/m², 554 g/m², 30 g/m² and 58.5 g/m² for nonmodified PB, TM, PP and TZ, respectively. Reduction in the absorptiveness of water was observed for impregnated PB with increase in gelatin content for all series of both hybrid systems. The smallest value was found for T3 sample of TEOS-based RGG5 systems (64 g/m²). In the case of TM the water absorptiveness decreased for all samples modified with all hybrid systems studied with increased gelatin content, similarly as it was in the case of PB. The lowest water absorptiveness (169 g/m²) was found for the last sample of FGG5 series based on U740. Contrary to previously tested paper samples, all samples of modified PP and TZ were characterized by increased absorptiveness of water. In the case of PP, the greatest increase in this parameter occurred for all T0 samples of all series based on TEOS (the relevant values: 44 g/m², 47 g/m², 44 g/m² and 47 g/m²) and for all U0 samples of all series of systems based on U740 (the relevant values: 37 g/m², 48 g/m², 37 g/m² and 48 g/m²). For all TZ samples the absorptiveness of water increased with the rise in gelatin content. The highest values of this parameter were found for last samples of all series for both hybrid systems. They were 195 g/m², 200 g/m², 182 g/m² and 182 g/m² for FGG5, FGG15, RGG5 and RGG15 series of TEOS-based systems and 167 g/m², 181 g/m², 190 g/m² and 198 g/m² for FGG5, FGG15, RGG5 and RGG15 series of U740-based systems.

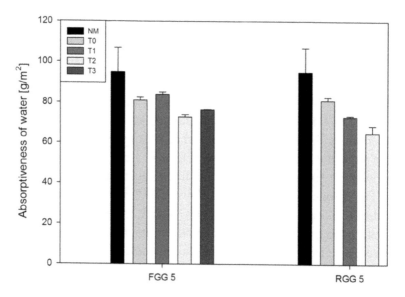

FIGURE 5.4 Absorptiveness of water of nonmodified (NM) PB and TEOS-modified (T0-T3) PB.

5.3.6 BURNING TIME

The measurement of this parameter consists in placing a paper sheet sample in the flame of a burner and determining the burning time (the burner was removed after the sample caught fire). Ten samples of each modified paper were subjected to the measurement and results were averaged. The averaged results of burning time measurements for PB modified with hybrid systems based on U740 and FGG as well as RGG are given in Fig. 5.5. Burning time for all series of modified paper samples was longer compared to nonimpregnated (NM) paper samples for which it was 9 s, 44 s, 9 s and 21 s for PB, TM, PP and TZ, respectively. In the case of PB the longest burning time (21 s) was observed for a sample from FGG5 series of TEOS-based system. Also in the case of TM an extension of burning time occurred with increasing gelatin content in all series of hybrid systems based on TEOS and U740. The longest burning time was 54 s for last samples from FG15 and RGG5 series of TEOS-based systems. An increase in burning time was also observed in the case of PP for which it was 17 s (for T3 sample from FGG5 series of TEOS-based systems) and in the case of TZ for which it was 35 s (for samples of TEOS-based systems).

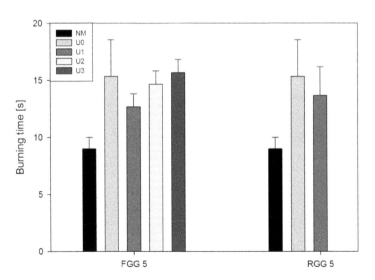

FIGURE 5.5   Burning time of nonmodified (NM) PB and PB modified with U740 (U0-U3).

5.4   CONCLUSIONS

Results of the study bring into conclusion that the kind of silica precursor (TEOS or ethyl silicate) has no significant effect on upgrading selected functional properties of paper products. More important parameters appeared to be the kind and amount of gelatin added. Such parameters as grammage, absorptiveness of water as determined by the Cobb method, bursting strength and burning time depend on gelatin content. A greater change in the studied parameters was observed in the case of RGG gelatin compared to food grade gelatin. It was also established that the hybrid materials applied as impregnates for paper products either considerably (™) or slightly (PB) improve functional parameters of materials nonmodified by manufacturers during the production process. In the case of paper products (PP, TZ) premodified in their production process, no changes or a slight deterioration of the studied parameters were observed.

KEYWORDS

- alkyl ketene dimers (AKDs)
- ethyl silicate (U740)
- modification
- properties of paper products
- silica—gelatin hybrid systems
- tetraethoxysilane (TEOS)

REFERENCES

1. Dupont, A.-L., J. Chromatogr. A 950 (2002) 113–124.
2. J. C. Roberts; Paper Chemistry, 2nd ed., Blackie, Glasgow, 1996.
3. Smitha, S., Shajesh, P., Mukundan, P., Nair, T., D. R., Warrier, K., G. K., Colloids and Surfaces B: Biointerfaces 55 (2007) 38–43.

4. Watzke, H. J., Dieschbourg, C., Adv. Colloid Interface Sci. 50 (1994) 1.
5. Schuleit, M., Luisi, P. L., Biotechnol. Bioeng. 72 (2001) 249.
6. Liu, D. M., Chen, I. W., Acta Mater. 47 (1999) 4535.
7. Yuan, G. L., Yin, M. Y., Jiang, T. T., Huang, M. Y., Jiang, Y. Y., J. Mol. Catal.: A, Chem. 159 (2000) 45.
8. Boanini, E., Rubini, K., Panzavolta, S., Bigi, A., Acta Biomaterialia, 6 (2010) 383–388.
9. Bigi, A., Cojazzi, G., Panzavolta, S., Roveri, N., Rubini, K., Biomaterials, 23 (2002) 4827–4832.
10. Karim, A. A., Bhat, R., Food Hydrocolloids, 23 (2009) 563–576.
11. Bigi, A., Cojazzi, G., Panzavolta, S., Rubini, K., Roveri, N., Biomaterials, 22 (2001) 763–768.
12. Ren, L., Tsuru, K., Hayakawa, S., Osaka, A., Biomaterials, 23 (2002) 4765–4773.
13. Advincula, M., Fan, X., Lemons, J., Advincula, R., Colloids Surf.:B, 42 (2005) 29.
14. Ren, L., Tsuru, K., Hayakawa, S., Osaka, A., Biomaterials, 23 (2002) 4765.
15. Ren, L., Tsuru, K., Hayakawa, S., Osaka, A., J. Non-Cryst. Solids, 285 (2001) 116.
16. Ren, L., Tsuru, K., Hayakawa, S., Osaka, A., J. Sol-Gel Sci.Tech., 21 (2001) 115.
17. Smitha, S., Mukundan, P., Pillai, P. K., Warrier, K., G. K., Mater. Chem. Phys., 103 (2007) 318.
18. Pietras, P., Przekop, R., Maciejewski, H., Ceramics Silikáty 57 (2013) 58–65.

CHAPTER 6

AN INFLUENCE OF A SIZE AND OF THE SIZE DISTRIBUTION OF SILVER NANOPARTICLES ON THEIR SURFACE PLASMON RESONANCE

A. R. KYTSYA, O. V. RESHETNYAK, L. I. BAZYLYAK, and YU. M. HRYNDA

CONTENTS

6.1   INTRODUCTION

On a basis of the comparative analysis of the references data the corre-
lated dependencies between the optical characteristics of aqueous sols of
spherical nanoparticles and their diameter have been discovered. As a result, the empirical dependencies between the values of the square of wave
frequency in the adsorption maximum of the surface plasmon resonance
and average diameter of the nanoparticles were determined as well as between the values of the adsorption band width on a half of its height and
silver nanoparticles distribution per size. Proposed dependencies are described by the linear equations with the correlation coefficients 0.97 and
0.84, correspondingly.

The field of nanoscience has blossomed over the last 20 years and the
need for nanotechnology will only increase, since the miniaturization becomes more important in such areas as computing, sensors and biomedical
applications. Advances in this field largely depend on the ability to synthesize nanoparticles of various materials, sizes and shapes as well as on
efficiency assemble them into the complex architectures. The early well–
known methods to produce suspensions of very small noble–metal particles are still used today and continue to be the standard by which other
synthesis methods are compared. The most popular method to synthesize
Au suspensions is the so-called Turkevich method, which employs the reduction of chloroauric acid with sodium citrate and produces a narrow
size distribution of 10 nm particles [1]. For Ag nanoparticles suspensions
a common method is the Lee–Meisel method, which is a variation of the
Turkevich method in that $AgNO_3$ is used as the metal source [2], but unlike
the Turkevich method, the Lee–Meisel method produces a broad distribution of particle sizes. The most common method for the synthesis of nano-
sized Ag particles is the reduction of $AgNO_3$ with $NaBH_4$. This method
can also be adapted to produce particles of other metals such as Pt, Pd,
Cu, Ni, etc. [3–6], although the specific protocols depend on the reduction
potential of the source ion. Cu and Ni suspensions, for example, are not
very stable since the metal particles are easily oxidized requiring strong
capping ligands to prevent the oxidation. Silver nanoparticles (Ag–NPs)
are characterized by unique combination of the important physical–chemical properties, namely by excellent optical characteristics, by ability to

amplify the signal in spectroscopy of the combination dispersion [7], and also by high antibacterial properties. Among the three metals (Ag, Au, Cu) that display surface plasmon resonances (SPR) in the visible spectrum, exactly Ag exhibits the highest efficiency of the plasmon excitation, that leads to the abnormally high value of the extinction coefficient of Ag–NPs [8]. Moreover, optical excitation of the plasmon resonances in nano-sized Ag–NPs is the most efficient mechanism by which light interacts with matter. A single Ag nanoparticle interacts with light more efficiently than a particle of the same dimension composed of any known organic or inorganic chromophore. Silver is also the only material whose plasmon resonance can be tuned to any wave-length in the visible spectrum. Under conditions of modern tendency to the miniaturization and the necessity to improve the technological processes of the new materials obtaining based on Ag–NPs, there is problem of their identification, which requests the cost equipment and causes a search of the alternative ways of their average size and of their size distribution determination by others methods, in particular, by calculated ones with the use of the empirical equations and dependencies which are based on the property of adsorption of the electromagnetic irradiation in UV/visible diapason by sols of Ag–NPs [9].

We have used the optical properties of silver, namely the dependence of SPR adsorption maximum position on a size of Ag–NPs as the characteristic of their size and the width of the adsorption band on a half of its height as Ag–NPs size distribution.

6.2 THEORETICAL GROUNDS

The SPR extinction spectra of Ag suspensions by different particle diameters are shown in Fig. 6.1 [10]. It is apparent that the dipole maximum rapidly shifts to longer wave-lengths as the particle size increases beyond 70 nm (450 nm spectral maximum) revealing the quadrupole peak at about 420 nm. The observed spectral shift results from the "spreading" of the particle's surface charge over a larger surface area so that the surrounding medium better compensates the restoring force thus slowing the electron oscillations [11].

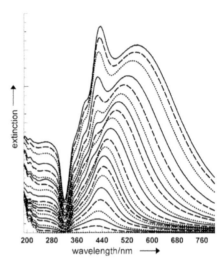

FIGURE 6.1 UV/Vis extinction spectra of silver nanoparticles suspensions for 20 different particle diameters [10].

A sufficiently small particle of any conducting material exhibits *SPR*s, yet its spectral position depends on many factors, most importantly on the material's frequency-dependent complex dielectric function. The wavelength dependence of the real $(\varepsilon_1(\omega))$ and imaginary $(\varepsilon_2(\omega))$ parts of the dielectric function describing polarizability and energy dissipation, respectively, are given in Fig. 6.2 for *Ag* [12].

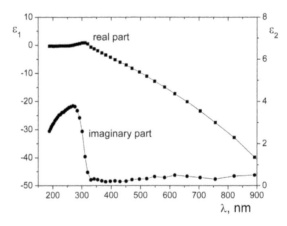

FIGURE 6.2 Real $(\varepsilon_1(\omega))$ and imaginary $(\varepsilon_2(\omega))$ parts of the dielectric function of silver as a function of wavelength. Curves were prepared using values listed in Ref. [12].

An SPR occurs when there is phase-matching between the polarization in the particle and incident field, a condition that is fulfilled for very small particles (<10 nm) when $\varepsilon_1(\omega) = -2\varepsilon_m$, where ε_m is the dielectric constant of the surrounding medium [13] and is satisfied for very small Ag particles suspended in water ($\varepsilon_m = 1.77$) at an excitation wavelength of around 385 nm. The imaginary part of the metal dielectric function, which describes losses, must be small at the SPR frequency to provide efficient electron oscillations. Several processes can dump the oscillations, such as electron scattering by lattice phonon modes, inelastic electron–electron interactions, scattering of the electrons at the particle surface, and excitation of bound electrons into the conduction band (interband transitions) [14]. Whereas electron–phonon interactions account for a majority of $\varepsilon_2(\omega)$, inelastic electron–electron interactions and surface scattering are less significant, with the latter being important only for <5 nm particles. Interband transitions can cause a substantially decreased efficiency of plasmon excitation as is the case for Au and Cu, where there is significant overlap between the interband adsorption edge and the plasmon resonance. For Ag, however, the adsorption edge is in the UV (320 nm) and has little impact on the SPRs, which appear at wavelengths larger than 370 nm, accounting for the fact that excitation of the SPR in Ag particles is more efficient than for Au and Cu.

For future practical applications of nanoparticles, synthesis techniques capable of producing the highly crystalline particles of many different sizes and narrow distribution are necessary as well as the determination of their size and of their size distribution.

6.3   RESULTS AND DISCUSSION

Generally, for the theoretical description of the SPR phenomenon of the metallic little particles and for the Ag–NPs, in particular, the solving's of the Maxell's equations are used, which in 1908 have been proposed by Gustav Mie [15]. Starting from the macroscopic Maxell equations, Gustav Mie calculated the extinction, scattering and absorption cross–sections of Au nanoparticles and showed how the spectra of the suspensions evolve as a function of particle size. The results of these calculations also allowed him to sketch scattering diagrams for different particle sizes and diagrams depicting the electric and magnetic fields of the dipole, quadrupole, octupole and sextupole components of the resonance. Now, it is well known that the optical resonances in noble–metal nanoparticles are the collective oscillations of conduction electrons termed "plasmons".

The extinction coefficient (C_{ext}) of the spherical nanoparticles in accordance with Mie's theory is described by the equation:

$$C_{ext} = \frac{24\pi^2 r \varepsilon_M^{3/2}}{\lambda} \frac{\varepsilon_2}{\left(\varepsilon_1 + 2\varepsilon_M\right)^2 + \varepsilon_2^2} \tag{1}$$

where r is the radius of a particle, λ is a length of a wave of the electromagnetic irradiation, ε_M is the dielectric transmissivity of the solvent, ε_1 is a real part of the value of dielectric transmissivity of a part of the metal, ε_2 is the imaginary part of the value of dielectric transmissivity of a part of the metal.

It is known [9, 10], that the position of SPR maximum adsorption depends on a size of the Ag–NPs. Such phenomenon is explained by dependence of real and imaginary parts of the dielectric permeability of silver on size of the nanoparticle. In accordance with Drude's model [16], ε_1 and ε_2 can be described by the expressions:

$$\varepsilon_1 = \varepsilon'_{bulk} + \frac{\omega_p^2}{\omega^2 + \omega_d^2} - \frac{\omega_p^2}{\omega^2 + \omega_r^2} \tag{2}$$

$$\varepsilon_2 = \varepsilon''_{bulk} + \frac{i\omega_p^2 \omega_r}{\omega\left(\omega^2 + \omega_r^2\right)} - \frac{i\omega_p^2 \omega_d}{\omega\left(\omega^2 + \omega_d^2\right)} \tag{3}$$

$$\omega_r = \omega_d + \frac{v_F}{r} \tag{4}$$

where ε'_{bulk} and ε''_{bulk} are values of the real and of the imagined parts of dielectric permeability of silver mass, ω, ω_p and ω_d are correspondingly the frequency of the electromagnetic irradiation, plasmon frequency of the metal and decrement of electron gas extinction in the mass metal, v_F is the Fermi rate.

However, calculated accordingly to such expressions adsorption spectra of aqueous sols of spherical Ag–NPs are differed from the experimental ones, that can be explained by different reasons, in particular: firstly, in presented example of the calculations it was not taken into account the distribution of Ag–NPs per sizes, that has an influence on a value of the SPR adsorption band width on a half of its height and, secondly, in classical Drude's model the adsorbed stabilizer on the surface doesn't take into account; in turn, such stabilizer can influence on the value of the wave length in adsorption maximum of the Ag–NPs sol.

In order to determine the dependencies between the optical characteristics and size of the nanoparticles we have done an analysis of the great data of Refs. [10, 17–52] concerning to the synthesis and the investigations of Ag–NPs.

It was determined (see Fig. 6.3), that a square of the wave frequency in adsorption maximum of SPR (ω^2) linearly depends on a value of the average diameter (d) Ag–NPs. Such dependence is described by the expression:

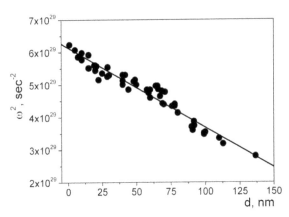

FIGURE 6.3 Dependence between the square of the wave frequency in adsorption maximum of SPR and diameter of Ag–NPs.

$$\omega^2 = (6,14 \pm 0,05) \cdot 10^{29} - (2,45 \pm 0,08) \cdot 10^{27} \tag{5}$$

with the correlation coefficient 0.97.

At the same time, it was not discovered the direct dependence between the width of the adsorption band of Ag–NPs on a half of its height ($\Delta\lambda$) and nanoparticles distribution per size (Δd). Evidently, it is connected with the nonmonotonic change of the adsorption band of Ag–NPs at their size increasing [16]. However, the all analyzed data are satisfactory described by the linear equation:

$$\log(d \cdot \Delta\lambda) = (0,2 \pm 0,1) + (0,89 \pm 0,06) \cdot \log(\Delta d \cdot \lambda_{max})$$ (6)

with the correlation coefficient 0.84 (*see* Fig. 6.4). Here λ_{max} is a value of the wavelength in a maximum of the SPR.

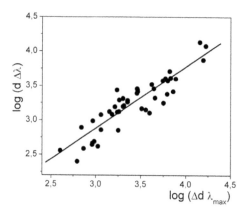

FIGURE 6.4 Dependence between the logarithms of compositions $\log(d \cdot \Delta\lambda)$ and $\log(\Delta d \cdot \lambda_{max})$ for Ag–NPs.

It is necessary to notify, that in processed data Ag–NPs were obtained in aqueous solution with the use of different upon nature stabilizers of the surface and precursors. However, in spite of this fact, discovered by us dependencies are good described with the respective correlation coefficients. It is clear, that for the explanation of nature for such dependencies the advanced theoretical analysis of the electron gas interaction with the electromagnetic irradiation is necessary, however, at the presented stage, such empirical dependencies can be used as the rapid method of the synthesized by different methods of Ag–NPs identification in laboratory and industrial conditions without the application of complicated, cost and often absent in Ukraine devices for their identification.

6.4 CONCLUSIONS

Empirical dependencies between the dimensional and optical character-istics of silver nanoparticles were determined. Such dependencies can be used for estimation of value of the average diameter and distribution per size of Ag–NPs without application of complicated equipment.

KEYWORDS

- optical characteristics
- plasmon resonance
- silver nanoparticles
- size distribution
- SPR phenomenon

REFERENCES

1. Turkevich, J., Stevenson, P. S., Hillier, J., *Disc. Faraday. Soc.* 11, 55 (1951).
2. Lee, P. C., Meisel, D. J., *Phys. Chem.* 86, 3391 (1982).
3. Scott, R. W. J., Ye, H., Henriquez, R. R., Crooks, R. M., *Chem. Mater.* 15, 3873 (2003).
4. Hou, Y., Gao, S., *J. Mater. Chem.* 13, 1510 (2003).
5. Sinha, A., Das Kumar, S., T. V. Vijaya Kumar, Rao, V., Ramachandrarao, P., *J. Mater. Synth. Proces.* 7, 373 (1999).
6. Sastry, M., Patil, V., Mayya, K. S., Paranjape, D. V., Singh, P., Sainkar, S. R., *Thin Solid Films* 324, 239 (1998).
7. Krutyakov, Y. A., Kudrinskiy, A. A., Olenin, A. Y., Lisichkin, G. V., *Russ. Chem. Rev.* 77 (3), 233 (2008).
8. Malynych, S. Z., *J. Phys. Stud.* 13 (1), 18011 (2009).
9. Malynych, S. Z., *J. Nano- Electron. Phys.* 2 (4), 5 (2010).
10. Evanoff, D. D., Chumanov, G., *Chem. Phys. Chem.* 6, 1221 (2005).
11. Kreibig, U., Vollmer, M., *Springer-Verlag, New York* (1995).
12. Johnson, P. B., Christy, R. W., *Phys. Rev. B.* 6, 4370 (1972).
13. Kawabata, A., Kubo, R., *J. Phys. Soc. Japan.* 21, 1765 (1966).
14. Ashcroft, N. W., Mermin, N. D., *Saunders College Publishing, Philadelphia* (1976).
15. Mie, G., *Ann. Phys.* 330 (3), 377 (1908).
16. Slistan-Grijalva, A., Herrera-Urbina, R., Rivas-Silva, J. F., Avalos-Borja, M., Castillo-Barraza, F. F., Posada-Amarillas, A., *Physica E.* 27, 104 (2005).

17. Kryukov, A. I., N. N. Zin'chuk, Korzhak, A. V., S.Ya. Kuchmii, *Theor. Exp. Chem.* 39 (1), 9 (2003).

18. Bryukhanov, V. V., Tikhomirova, N. S., Gorlov, R. V., V.A. Slezhkin, *Izviestiya KGTU.* 23, 11 (2011).

19. Munro, C. H., Smith, W. E., Garner, M., Clarkson, J., White, P. C., *Langmuir.* 11, 3712 (1995).

20. Podliegayeva, L. N., Russakov, D. M., Sozinov, S. A., Morozova, T. V., Shvajko, I. L., Zvidentsova, N. S., Koliesnikova, L. V., *Viestnik KGU.* 2 (38), 91 (2009).

21. Parashar, V., Parashar, R., Sharma, B., Pandey, A., *Dig. J. Nanomater. Bios.* 4 (1), 45 (2009).

22. Yin, Y., Z.Yu. Li, Zhong, Z., Gates, B., Xia, Y., Venkateswaranc, S., *J. Mater. Chem.* 12, 522 (2002).

23. Sun, Y., Xia, Y., *Analyst.* 128, 686 (2003).

24. Wang, H., Qiao, X., Chen, J., Ding, S., *Colloid. Surface A.* 256, 111 (2005).

25. Mohan, Y. M., Lee, K., Premkumar, T., Geckeler, K. E., *Polymer.* 48, 158 (2007).

26. Li, X., Zhang, J., Xu, W., Jia, H., Wang, X., Yang, B., Zhao, B., Li, B., Yu. Ozak, *Langmuir.* 19 (10), 4285 (2003).

27. Vigneshwaran, N., Nachane, R. P., Balasubramanya, R. H., Varadarajan, P. V., *Carbohydr. Res.* 341, 2012 (2006).

28. Shankar, S., Rai, A., Ahmad, A., Sastry, M., *J. Colloid. Interf. Sci.* 275, 496 (2004).

29. Chandran, S., Chaudhary, M., Pasricha, R., Ahmad, A., Sastry, M., *Biotechnol. Prog.* 22, 577 (2006).

30. Hiramatsu, H., Osterloh, F., *Chem. Mater.* 16 (13), 2509 (2004).

31. Song, J. Y., Kim, B., *Bioproc. Biosyst. Eng.* 32, 79 (2009).

32. Li, S., Yu. Shen, Xie, A., Yu, X., Qiu, L., Zhang, L., Zhang, Q., *Green. Chem.* 9, 852 (2007).

33. Zhu, J., Liu, S., Palchik, O., Yu. Koltypin, Gedanken, A., *Langmuir.* 16, 6396 (2000).

34. Shahverdi, A. R., Minaeian, S., Shahverdi, H. R., Jamalifar, H., Nohi, A. A., *Process. Biochem.* 42, 919 (2007).

35. Martinez-Castanon, G. A., N. Nino–Martinez, Martinez-Gutierrez, F., Martinez-Mendoza, J. R., Ruiz, F., *J. Nanopart. Res.* 10, 1343(2008).

36. Shahverdi, A., Fakhimi, A., Shahverdi, H., Minaian, S., *Nanomed-Nanotechnol.* 3 (2), 168 (2007).

37. Huang, H., Yang, X., *Carbohydr. Res.* 339 (15), 2627 (2004).

38. Ahmad, A., Mukherjee, P., Senapati, S., Mandal, D., Khan, M. I., Kumar, R., Sastry, M., *Colloid. Surface B.* 28 (4), 313 (2003).

39. Nickel, U., Mansyreff, K., Schneider, S., *J. Raman Spectrosc.* 35, 101 (2004).

40. Steven, J. O., *Sigma-Aldrich Co.* http:,www.sigmaaldrich.com/materials-science/nanomaterials/silver-nanoparticles.html

41. Huang, H. H., Ni, X. P., Loy, G. L., Chew, C. H., Tan, K. L., Loh, F. C., Deng, J. F., Xu, G. Q., *Langmuir* 12, 909 (1996).

42. Henglein, A., Giersig, M., *J. Phys. Chem. B.* 103, 9533 (1999).

43. http:,www.nanocomposix.com/products/silver/spheres

44. O. V. Dement'eva, A. V. Mal'kovskii, Filippenko, M. A., Rudoy, V. M., *Colloid J+.* 70 (5), 561 (2008).

45. Sharma, V., Yngard, R., Lin, Y., *Adv. Colloid Interfac.* 145, 83 (2009).
46. Tilaki, R. M., Irajizad, A., Mahdavi, S. M., *Appl. Phys. A* 4, 215 (2006).
47. Panacek, A., Kvitek, L., Prucek, R., Kolar, M., Vecerova, R., Pizurova, N., Sharma, V. K., Nevecna, T., Zboril, R., (2006) *J. Phys. Chem. B.* 110:16248–16253.
48. Tien, D. C., Liao, C. Y., Huang, J. C., Tseng, K. H., Lung, J. K., Tsung, T. T., Kao, W. S., Tsai, T. H., Cheng, T. W., Yu, B. S., Lin, H. M., Stobinski, L., *Rev. Adv. Mater. Sci.* 18, 750 (2008).
49. Liu, J., Hurt, R., *Environ. Sci. Technol.* 44, 2169 (2010).
50. D. L.Van Hyning, Zukoski, C. F., *Langmuir.* 14, 7034 (1998).
51. Solomon, S. D., Bahadory, M., Jeyarajasingam, A. V., Rutkowsky, S. A., Boritz, C., *J. Chem. Educ.* 84 (2), 322 (2007).
52. Abkhalimov, E. V., Parsaev, A. A., Ershov, B. G., *Colloid J+.* 73 (1), 1 (2011).

CHAPTER 7

AN INFLUENCE OF THE KINETIC PARAMETERS OF THE REACTION ON A SIZE OF OBTAINED NANOPARTICLES AT THE REDUCTION OF SILVER IONS BY HYDRAZINE

A. R. KYTSYA, YU. M. HRYNDA, L. I. BAZYLYAK, and G. E. ZAIKOV

CONTENTS

7.1   INTRODUCTION

Kinetics of the reaction of silver nitrate reduction by hydrazine in the presence of sodium citrate in alkaline medium was studied in a wide range of the reagents concentration variation. The orders of the reaction were determined and the effective constants of the silver nanoparticles nucleation process rate and of their propagation one were calculated. It was investigated the optical characteristics of the obtained sols of the silver nanoparticles. It was determined the empirical dependence of a size of the obtained silver nanoparticles on the kinetic parameters of a process.

An exponential growth in a field of the fundamental and applied sciences connected with a synthesis of the nanoparticles of noble metals, studies of their properties and practical application is observed for the last decades. Silver nanoparticles *(Ag–NPs)* are characterized by unique combination of the important physical–chemical properties, *namely* by excellent optical characteristics caused by the phenomena of the surface plasmon resonance [1], by ability to amplify the signal in spectroscopy of the combination dispersion and also by high antibacterial properties. Despite the fact, that there are a number of methods for the synthesis of different upon nature nanoparticles and nanomaterials describing in Refs. [2–7], however the kinetic peculiarities and regularities of the formation (nucleation and propagation) of nanoparticles studied insufficiently and have the episodical character [8–10].

That is why the aim of the presented work was to investigate an influence of the synthesis conditions on the kinetic parameters of a process and also on the form and on the size of the synthesized *Ag–NPs*.

7.2   EXPERIMENTAL SECTION

Ag–NPs were obtained in accordance with the reaction (1) via reduction of silver nitrate by hydrazine in aqueous medium in the presence of sodium hydroxide at 20°C:

$$4\,AgNO_3 + 4\,NaOH + N_2H_4 = 4\,Ag + 4\,NaNO_3 + 4\,H_2O + N_2\uparrow \qquad (1)$$

Sodium citrate has been used as the stabilizer at its concentration 1×10^{-4} mol/L in the reactive solution.

Kinetic of the reaction was studied by direct potentiometric method with the use of the ion-selective microelectrode "*ELIS*–131 Silver". The concentration of the silver ions was determined continuously during the reaction proceeding per change of the potential of the ion-selective electrode regarding to the chlorine–silver comparison electrode. In order to avoid the hit of the chlorine ions into the reactive mix the salt weak link with the potassium nitrate was used.

The form and the average diameter of the silver nanoparticles were estimated with the use of the scanning electron microscopy *EVO*–40*XVP (Carl Zeiss)* with a system of the *X*-ray microanalysis *INCA* Energy, *XRD*-analysis and also on a basis of the adsorption spectra of the surface plasmon resonance of *Ag–NPs* sols with the use of the single-beam spectrophotometer UV-visible range *UV-mini*-1240 (*P/N* 206-89175-92; *P/N* 206-89175-38; *Shimadzu Corp., Kyoto, Japan*).

XRD-analysis was carried out with the use of *XRD* diffractometer *DRON*-3.0 with Cu−K$_\alpha$ irradiation (λ = 1,5405 *nm*). Data have been analyzed by full-profile revision in accordance with the *Ritveld's* method with the use of the simulation package *GSAS (General Structure Analysis System)*.

7.3   RESULTS AND DISCUSSION

In order to explain and to investigate the chemical process of the silver nanoparticles synthesis the kinetic characteristics have been investigated, namely the change of the concentration of the silver ions during the experiment. Typical kinetic curves of the silver concentration change via time are represented on Fig. 7.1.

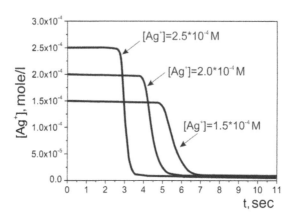

FIGURE 7.1 Kinetic curves of the reduction reaction of silver ions by hydrazine at different starting concentrations of the silver nitrate.

As we can see, the starting section of the kinetic curve corresponds to the stage of the nucleus centers formation and the following sharp decrease of the concentration of silver ions corresponds to their growth stage. Therefore, the time (t_0) of the initial section of the kinetic curve determines the Ag–NPs nucleation rate $(W_0 = 1/t_0)$, and the tangent of the angle of inclination of its quick linear section determines the rate of the nucleus growth (W_{max}).

It was investigated an influence of the change of starting concentrations of silver nitrate, sodium hydroxide and hydrazine on kinetic parameters of a process. In order to increase the trustworthiness of the results, the series of investigations consisting of 7–10 experiments has been carried out for every concentration. Obtained data were averaged. The ratio error at the W_0 and W_{max} determination not exceeds 25%.

Taking into account, that the reagents change concentration at the nucleation stage and also at the stage of the starting section of the silver NPs growth is insignificant, it can be used with some approximation the dependencies of the rates of a process on the starting concentrations of reactive mix components in order to determine the orders of the reaction.

Presented on Fig. 7.2 data indicate that the orders of the reaction per every among reagents for the processes of silver nanoparticles nucleation and growth are agreed and are equal to 1, 1 and 1/2 for $AgNO_3$, $NaOH$ and N_2H_4 correspondingly.

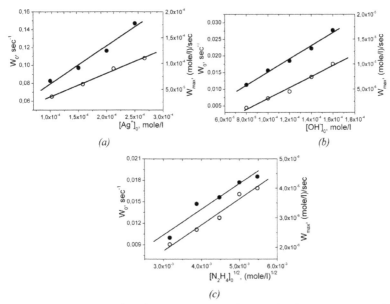

FIGURE 7.2 **FIGURE 7.2** Dependencies of the silver nanoparticles nucleation rate W_0 (●) and of the growth rate W_{max} (○) on the initial concentrations of $AgNO_3$ (a), $NaOH$ (b) and N_2H_4 (c).

Taking into account the determined orders of the reactions per every components, obtained experimental results can be represented in the coordinates of the acting masses law (2):

$$W = k^{\#} \, [Ag^+]^1 \times [OH^-]^1 \times [N_2H_4]^{1/2} \tag{2}$$

and the numerical values of the effective constants rates of the processes of new phase nucleation ($k_f^{\#} = (2.2 \pm 0.2) \times 10^8$ $(mol/L)^{-2.5} \times s^{-1}$) and its growing ($k_g^{\#} = (1.8 \pm 0.4) \times 10^5$ $(mol/L)^{-2} \times s^{-1}$) can be estimated (Fig. 7.3).

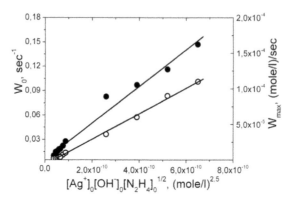

FIGURE 7.3 Dependencies of W_0 (●) and W_{max} (○) on composition of the initial concentrations of the reagents in coordinates of the Eq. (2).

In order to identify the obtained silver nanoparticles, their spectral characteristics were investigated (Fig. 6.4a). The all spectra of silver nanoparticles adsorption are characterized by one maximum corresponding to their spherical form [5]. Analyzing the references, it was discovered that the value of the square of wave frequency in adsorption maximum of the surface plasmon resonance of silver nanoparticles linearly depends on their size (Fig. 6.4b) that gives the possibilities to calculate an average diameter of the obtained silver nanoparticles. Calculated values of the average diameter of silver nanoparticles consist of 13–35 nm depending on the synthesis conditions.

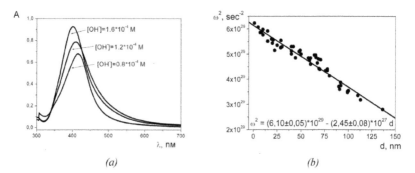

FIGURE 7.4 Adsorption electron spectra of silver nanoparticles obtained at different starting concentrations of sodium hydroxide (a) and calibration plot for calculation of their average diameter (b) based on the Refs. [6–7, 11–15].

Silver nanoparticles obtained at 20°C and starting concentrations of reagents $[AgNO_3]_0 = 2.5 \times 10^{-4}$ M, $[NaOH]_0 = 3.0 \times 10^{-4}$ M, $[N_2H_4]_0 = 7.5 \times 10^{-5}$ M were investigated with the use of the *SEM* and *XRD* analysis methods for confirmation of the obtained calculations (Fig. 5). On a basis of the *XRD*-analysis results, an average size of the silver crystallites was calculated; it consists of $D_V = 9.3$ nm, respectively, the diameter of the spherical particle for monodisperse system is $D = 4/3$ and $D_V = 12.4$ nm. Calculated accordingly to the location of the surface plasmon resonance adsorption maximum of the sol silver nanoparticles (Fig. 4b) value of average diameter of silver nanoparticles obtained under such conditions consists of 13 nm.

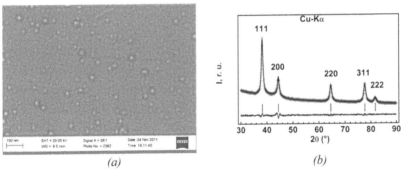

(a)   (b)

FIGURE 7.5   *SEM* image (a) and XRD-spectrum (b) of silver nanoparticles obtained at the starting concentrations $[AgNO_3]_0 = 2.5 \times 10^{-4}$ mol/L, $[NaOH]_0 = 3.0 \times 10^{-4}$ mol/L, $[N_2H_4]_0 = 7.5 \times 10^{-5}$ mol/L.

At the analysis of the experimental data it was found that the size of the obtained silver nanoparticles depends on the ratio of the nuclear centers formation rate and the nuclear centers growth rate (Fig. 7.6).

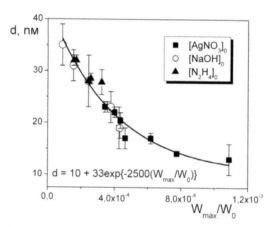

FIGURE 7.6 Dependence of an average diameter of silver nanoparticles on kinetic parameters of the process: an average diameter of silver nanoparticles obtained at different starting concentrations of $AgNO_3$ (■), $NaOH$ (○) and N_2H_4 (▲).

Such dependence can be explained by fact that at the nuclear centers formation rate increasing not only the concentration of the nuclear centers is increased, but also their critical radius is decreased that leads to the decreasing of the average size of synthesized nanoparticles. Evidently, such dependence is the partial case and is determined by the conditions of the synthesis and by the choice of the reagents, however permits to obtain the silver nanoparticles by the controlled size.

7.4   CONCLUSIONS

The reaction of the silver ions reduction by hydrazine has been investigated in alkaline medium in the presence of the sodium citrate. It was determined that the rates of the nucleation and of the growth of silver nanoparticles linearly depend on the initial concentrations of silver nitrate, sodium hydroxide and on concentration of hydrazine in degree 1/2. It was discovered the empirical dependence of a size of obtaining nanoparticles on the ratio of the rates of growth and nucleation of silver nanoparticles.

ACKNOWLEDGMENT

This work was performed via the framework of the Task Complex Program TCPSI "RESURS" (grant # P 8.2–2013/K).

KEYWORDS

- **hydrazine**
- **influence**
- **kinetic parameters**
- **nanoparticles**
- **reaction**
- **reduction of silver ion**

REFERENCES

1. Henglein, A. Small-Particle Research: Physicochemical Properties of Extremely Small Colloidal Metal and Semiconductor Particles. *Chem. Rev.* 1989, *89* (8), 1861–1873.
2. Suzdalyev, I. P. Nanotechnology: physicochemistry of nanoclaters, nanostructures and nanomaterials. M.: "KomKniga", 2006, 592 p. (in Russian)
3. Pomogaylo A. D., Rozenberg A. S., Uflyand A. S. Nanoparticles of metals in polymers. M.: "Khimiya", 2002, 672 p. (in Russian)
4. Egorova E. M., Revina A. A. ect. Bactericidal and catalytic properties of stable metallic nanoparticles in reverse micelles *Vestn. Mosc. Univ. Ser. 2. Khimiya.* 2001, *42* (5) 332–338 *(in Russian)*.
5. Krutyakov, Y. A., Kudrinskiy, A. A., Olenin, A. Y., Lisichkin, G. V. Synthesis and properties of silver nanoparticles: advances and prospects. *Russ.Chem.Rev. 2008, 77*(3), 233–258.
6. David, D. Evanoff Jr., Chumanov, G. Synthesis and Optical Properties of Silver Nanoparticles and Arrays. *Chem. Phys. Chem.*, 2005, *6*, 1221–1231.
7. Kryukov, A. I., Zin'chuk, N. N., Korzhak, A. V., Kuchmii, S.Ya. The effect of the conditions of catalytic synthesis of nanoparticles of metallic silver on their plasmon resonance. *Theor. Exp. Chem.* 2003, *39* (1), 9–14.
8. Khan, Z., Al-Thabaiti, S. A., El-Mossalamy, E. H., Obaid, A. Y. Studies on the kinetics of growth of silver nanoparticles in different surfactant solutions. *Colloids and Surfaces B: Biointerfaces.* 2009, *73*, 284–288.

9. Chou, K.S, Lu, Y. C., Lee, H. H. Effect of alkaline ion on the mechanism and kinetics of chemical reduction of silver. *Materials Chemistry and Physics*. 2005, *94*, 429–433.

10. Ignatyev A. I., Nashchyokin A. V., Nieviedomsky, V. M., Podsvirov O. A., Sidorov A. I., Solovyov A. P., Usov O. A. Peculiarities of silver nanoparticles formation in photo-thermorefractive glasses at electron irradiation. *Journal of technical physics*. 2011, *81* (5), 75–80 (in Russian).

11. http:,www.sigmaaldrich.com/materials-science/nanomaterials/silverna-noparticles. html (*information about products*).

12. Bryukhanov, V. V., Tikhomirova, N. S., Gorlov, R. V., Sliezhkin, V. A. An interaction of surface plasmons of silver nanoparticles on silokhrome with electron excited adsorbates of molecules of rhodamine 6Zh. *Izviestiya KGTU*. 2011, *23*, 11–17 (in Russian).

13. Munro, C. H., Smith, W. E., Garner, M., Clarkson, J., White, P. C. Characterization of the Surface of a Citrate-Reduced Colloid Optimized for Use as a Substrate for Surface-Enhanced Resonance Raman Scattering. *Langmuir*. 1995, *11*, 3712–3720.

14. Podlyegayeva, L. N., Russakov, D. M., Sozinov, S. A., Morozova T. V., Shvaiko, I. Л., Zvidentsova, N. S., Kolesnikov, L. V. An investigation of properties of silver nanoparticles obtained by the reduction from the solutions and by thermal sputtering in vacuum. *Vestnik Kemerovskiego gossudarstvennogo universiteta*. 2009, *2* (38), 91–96 (in Russian)

15. http://nanocomposix.com/products/silver/spheres (*information about products*).

CHAPTER 8

KINETICS AND MECHANISM OF INTERACTION BETWEEN OZONE AND RUBBERS

V. V. PODMASTERYEV, S. D. RAZUMOVSKY, and G. E. ZAIKOV

CONTENTS

8.1   INTRODUCTION

In this chapter, the existing points of view on the mechanism of protective action of antiozonant are discussed. The interrelation between protective properties of the studied connections and value of a constant of speed of reaction ozone-antiozonant is shown.

Destruction of nonsaturated polymers (mainly—rubbers and rubbers) under the influence of ozone is accompanied by formation of cracks on a surface of products, loss of mechanical durability and destruction. In literature there is enough of researches and some survey works, which are well generalizing the main results of researches of ozonic destruction of rubbers and ways of fight against it [1–4]. However, despite a large number of researches yet it wasn't succeeded to achieve effective protection against ozonic destruction. Works on search of new protective systems proceed [5, 6]. In the most part these works have empirical character.

8.2   EXPERIMENTAL RESULTS AND DISCUSSION

The technique of synthesis of ozone, its registration, calculations of constants of speeds of reaction are given in Ref. [7].

Most of the highly effective antiozonants belong to the group if substituted N, N'-phenylenediamines (PPD) and perform a great number of diverse functions. Apart from protecting against atmospheric ozone these compounds serve as effective antioxidants and antifatigue agents.

Figure 8.1 summarizes the results of dynamic testing of natural rubber protected by various antiozonants [8]. Plotted on the ordinate is the mean crack area (S) per unit area of the specimen, which was used in this study as a measure of degradation:

$$S = \sum_{i=1}^{i=n} \frac{lh}{n} \tag{1}$$

where n is the number of cracks, l is the length of a crack, and h is its depth.

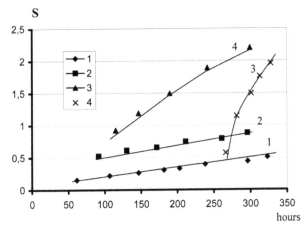

FIGURE 8.1   Effect of test time on the degree of degradation (*S*) of the surface of naturally derived rubber containing various antiozonants: 1, *N*, *N'*-di-1-methylheptil-p-phenylenediamine; 2, N-phenyl-N'isopropyl-p-phenylenediamine; 3, 6-ethoxy-1,2-dihydro-2,2,4-trimetylquinoline; 4, N, N'-dioctil-p-phenylenediamine.

Most of the other properties of the vulcanizes also improve perceptibly. Antiozonants increase the time to cracking, improve the specimen durability [9, 10], minimize creep, and reduce the stress relaxation rate in the specimen [11].

The complexity of the cracking phenomena is matched by that of antiozonant action. Several proposals have been made regarding the mechanism of this protective action in polymers; however, there is only one work in which all possible mechanisms are compared within a single experiment [12].

The suggestion that antiozonants serve a catalysts of ozone decomposition [13] cannot be true because ozone react with the antiozonant [14], and the stoichiometry and main products of the reaction are known [15, 16].

Not truth an accept as the proposal [17, 18] that, being bifunctional, antiozonats react with products of interaction between ozone and macromolecules, binding the ends of the disrupted chains, since it is clear that the important objective is not to heal broken chains but to maintain the existing sequences of bonds. Furthermore, monofunctional antiozonants such as tributylthioureas [19] are known which cannot broken ands but are capable of adequately protecting rubbers against atmospheric ozone.

Another hypothesis, which states that antiozonants migrate to the surface where they react with ozone itself or with the products of its reaction with the C=C bonds of the polymer thus forming a film that prevents ozone from penetrating inside toward the polymer, is difficult to verify. A similar effect is to be expected by analogy with waxes. However, there are two indirect arguments against any significant effect of film formation. Firstly, the products of the reaction between ozone and N-isopropyl-N'-phenyl-p-phenylenediamine (4010NA), incorporated into a specimen of cured natural rubber, do not protect it against ozone-induced degradation [16]. Secondly, calculation of the antiozonant diffusion rate indicates that migration toward the surface from the core of the specimen cannot play any significant role during cracking [24].

Nevertheless, judging from many publications [17, 18, 20–23] most researchers believe that the physical properties of antiozonants play an important role in the efficiency of their protective action, although it is not specified which properties. In particular, it seems that only the difference in the physical properties of tributylurea and 4010NA can explain why the times to cracking are comparable only under static conditions, while under dynamic conditions tributylurea is much less effective [27]. Comparison of the behavior of tributylthioureas with different substituents suggest that alkyl substituents with an open chain are replaced by cyclic ones, the correlation between the time to cracking and reactivity to ozone remains the same although the curves diverge further (*see* Fig. 8.2) [19].

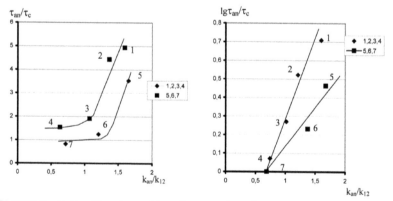

FIGURE 8.2 Ratio of time to cracking of protected specimens (τ_{an}) to that of unprotected control specimen (τ_c) versus ratio between rate constants of the ozone-antiozonant (k_{an}) and ozone-C=C (k_{12}) reactions.

The proposal [17] that antiozonants on the rubber surface compete with the C=C bonds for ozone and thereby protect vulcanisates against cracking is adequately corroborated at present. The arguments supporting this assumption have been derived using kinetic methods for studying ozone-macromolecule and ozone-antiozonant reactions insolutions [15, 16, 25]. It was shown that in the presence of antiozonants the rate of macromolecular chain scission slowed down (Fig. 8.3) although the absorption of ozone continued. Products of the reaction of ozone with antiozonant accumulated in the solution. Studies into the kinetics of this reaction have shown that it is extremely fast, its rate exceeding those of all presently known reactions involving ozone [26].

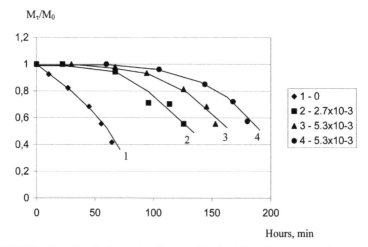

FIGURE 8.3   Variation in the molecular weight of polybutadiene exposed to ozone in a solution at various concentrations of N-phenyl-N'-isopropyl-p-phenilene diamine (mol/L): 1, 0; 2, 2.7×10^{-3}; 3 and 4, 5.3×10^{-3}. Molecular weight of polybutadiene: 1–3, 60,000; 4, 4000, CCl_4, $[O_3] = 4 \times 10^{-5}$ mol/L, 20°C.

1. C_4H_9NH —— C —N$(C_4H_9)_2$;   2. ⬡ NH — C —N$(C_4H_9)_2$; (K_{12})
                      ‖   ‖
                      S   S

3. 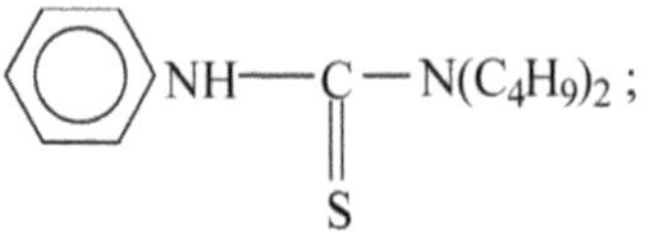 $\langle\bigcirc\rangle$NH—C—N(C$_4$H$_9$)$_2$;   4. O$_2$N$\langle\bigcirc\rangle$NH—C—N(C$_4$H$_9$)$_2$;
                    ‖   ‖
                    S   S

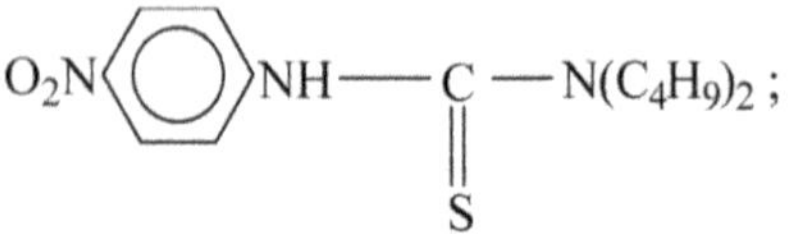

5. 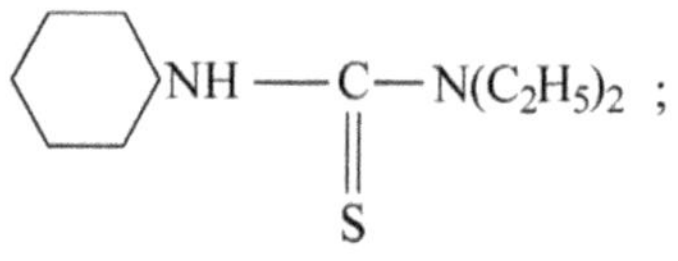 $\langle\bigcirc\rangle$NH—C—N(C$_2$H$_5$)$_2$;   6. 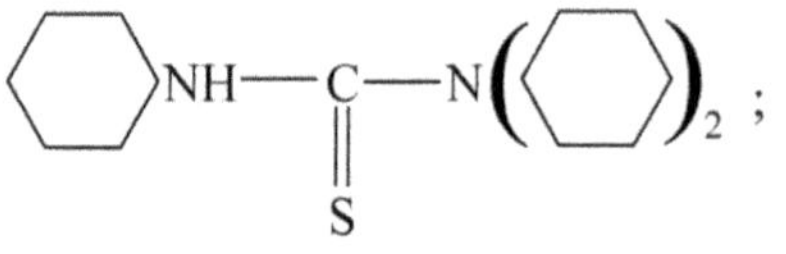 $\langle\bigcirc\rangle$NH—C—N$\left(\langle\bigcirc\rangle\right)_2$;
                    ‖   ‖
                    S   S

                              C$_2$H$_5$
                               |
7. 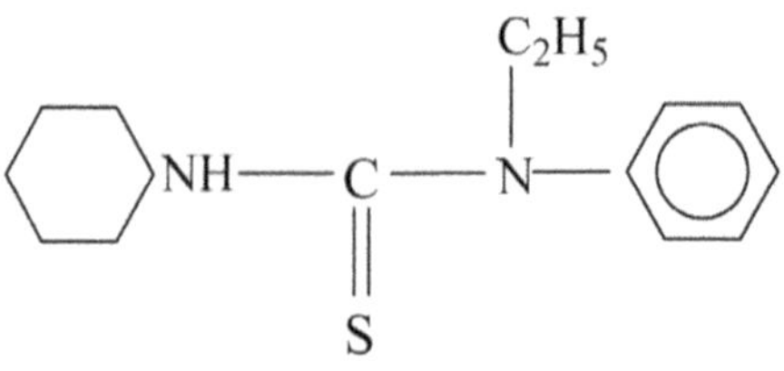 $\langle\bigcirc\rangle$NH—C—N—$\langle\bigcirc\rangle$
                    ‖
                    S

Special methods have been developed for measuring the rate constants of the ozone-antiozonant reaction. One method is based on the competition of ozone for a reference compound with a known rate constant and the compound of interest [27], while the other is based on comparison on the ozone dissolution rate and the rate of its reaction [28, 29].

In the first method,[27] a thiourea which reacts with ozone in two steps (Scheme 1) was selected as the reference compound:

Scheme 1

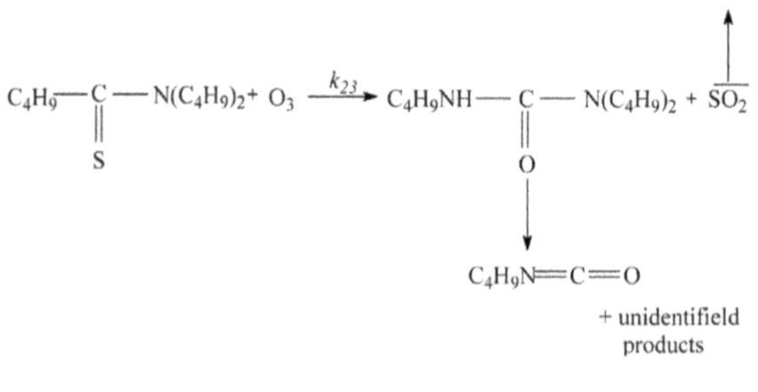

$$C_4H_9\underset{S}{\overset{\|}{-C-}}N(C_4H_9)_2 + O_3 \xrightarrow{k_{23}} C_4H_9NH-\underset{O}{\overset{\|}{C}}-N(C_4H_9)_2 + \uparrow SO_2$$

$$\downarrow$$

$$C_4H_9N{=}C{=}O$$

+ unidentifield
products

The first step yields sulfur dioxide. When the reaction is conducted in a bubble reactor, the presence of SO$_2$ can be detected by absorption in the UV region.

Figure 8.4 represents typical curves for variation in the optical density (D) of the mixture at the reactor outlet for the interaction of reference compound, methyl oleate, and tributylthiourea (TBTU) with ozone. Knowing the starting amounts of methyl oleate and TBTU, the gas flow rate and O_3 concentration one can easily calculate the stoichiometric coefficients of the reactions using the area above the curve before ozone appears at the reactor outlet. These were found to be 1 for methyl oleate and 2 for tributylthiourea.

Comparison of the curves representing the optical density of the gaseous mixture at the reactor outlet versus time indicates that the change in optical density at $\lambda = 280–300$ nm (Fig. 8.4b) differs in shape from that observed at 254 nm, which is representative of the kinetics of the variation in the absorption in the ozone rate (Fig. 8.4a). The first step of the reaction between TBTU and ozone yields SO_2 whose absorption maximum can be seen at 285 nm, and this accounts for the increase in optical density during the course of the reaction.

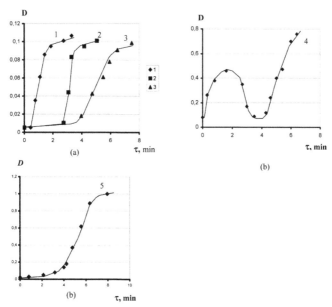

FIGURE 8.4 Optical density (**D**) of gaseous mixture at the reactor outlet versus time. (a) $[O_3]_0 = 6 \times 10^{-6}$ mol/L; $\lambda = 254$ nm; $V_{O2} = 120$ mL/min; 1, CCl_4; 2, 6×10^{-5} mol/L of methyl oleate in CCl_4; 3, 6×10^{-5} mol/L of tributylthiourea in CCl_4; (b) $[O_3]_0 = 2 \times 10^{-3}$ mol/L; $\lambda = 290$ nm; $V_{O2} = 120$ mL/min; 4, 6×10^{-3} mol/L of tributylthiourea in CCl_4; 5, 5×10^{-3} mol/L of methyl oleate in CCl_4.

The resulting SO_2 accumulates in the solution but is partially carried away by the gas flow (Scheme 2).

Scheme 2

$$[O_3]_{sol} + TBTU \xrightarrow{k_{21}} [products] + [SO_2]_{sol}$$

$$[SO_3]_{sol} \underset{}{\overset{\alpha}{\rightleftharpoons}} SO_{2gas}$$

where α is the solubility coefficient. In the case the rate of the reaction between TBTU and ozone and the rate of formation of the volatile component depend on the gaseous mixture supply rate and are constant with time ($W_p = \omega[O_3]_0$), where ω is the specific rate of ozone supply ($L^{-1}s^{-1}$), hence

$$\omega[O_3]_0 \alpha\tau = W[SO_2]_{gas}\alpha\tau + [SO_2]_{sol} \tag{2}$$

or
$$\frac{\alpha[SO_2]_{sol}}{\alpha\tau} = \omega([O_3]_0 - [SO_2]_{gas}) \tag{3}$$

Substitution of $\alpha[SO_2]_{gas}$ for $\alpha[SO_2]_{sol}$ and integration gives Eq. (4):

$$\omega[SO_2]_{gas} = \omega[O_3]_0\left[1 - \exp\left(-\frac{\omega\tau}{\alpha}\right)\right] \tag{4}$$

where $[SO_2]_{gas}$ is the rate of release of volatile products, $\omega[O_3]_0$ is the rate of the reaction between ozone and TBTU, and $[1-\exp(\omega\tau/\alpha)]$ is the reaction factor of SO_2 in the solution.

When olefin is introduced into the system, the ozone reacts with both components at the same time (Eq. (5)):

$$\omega[O_3]_{gas} = k_{21}[TBTU][O_3] + k_{12}[olefin][O_3] \tag{5}$$

and the participation of TBTU in the total process is given by the Eq. (6)

$$\frac{k_{21}[TBTU]}{k_{21}[TBTU] + k_{21}[olefin]} \tag{6}$$

then

$$\omega[SO_2]_{gas} = \omega[O_3]_0 \frac{k_{21}[TBTU]}{k_{21}[TBTU]+k_{12}[olefin]}\left[1-\exp\left(-\frac{\omega\tau}{\alpha}\right)\right]$$ (7)

The SO_2 concentration at the reactor outlet will diminish accordingly. The dependence of $[SO_2]_{gas}$ on [olefin] is shown in Fig. 8.5. If one uses the ratio $[SO_2]'_{gas}/[SO_2]''$ (D' and D" are the corresponding optical densities) at different olefin concentrations [olefin]' and [olefin]" within the same period of time, one can easily determine k_{21} if k_{12}, the rate constant of the reaction between ozone and olefin, is known.

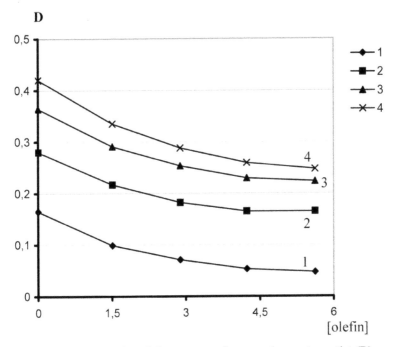

FIGURE 8.5 Optical density of the gaseous mixture at the reactor outlet (D) versus olefin concentration [olefin] at various times: 1, 1 min; 2, 2 min; 3, 3 min; 4, 4 min (Tributylthiourea, 4.1×10^{-4} mol/L; $[olefin]_0 = 1.5 \times 10^{-3}$ mol/L; $V_{O_3} = 120$ mL/min; $\lambda = 290$ nm).

$$k_{21} = \frac{k_{12}[\text{olefin}]'' - \dfrac{D'}{D''}[\text{olefin}]'}{[\text{TBTU}]\left(\dfrac{D'}{D''} - 1\right)} \tag{8}$$

In the case of tributylthiourea the rate constant of the reaction with ozone (k_{21}) was found to be $(2\pm0.2) \times 10^6$ L/mol s.

The rate constants of the reaction of ozone with reference olefin $(k_{21} = 1 \times 10^6$ L/mol s) and with the product of the first step of the reaction between ozone and TBTU isocyanate $(k_{22} = (3.5 \pm 0.5) \times 10^3$ L/mol s) were calculated from the slopes of curves 2 and 3 in Fig. 8.4(a) using the formula:

$$k_{22} = \frac{\omega[O_3]_0 - [O_3]_{gas}}{\alpha[X]\tau[O_3]_{gas}} = \frac{(D_0 - D_\tau)}{\alpha[X]_\tau D_\tau} \tag{9}$$

where $[X]_\tau$ is the current concentration of isocyanate (or methyl oleate with ozone), while D_0 and D_τ are the optical densities of gaseous mixture at the system inlet and outlet initially.

These data suggest that the rate constant of the reaction between ozone and TBTU is much greater than those of the reactions of ozone with the C=C bonds in polymers $(k_3 = 1\cdot4 \times 10^5$ L/mol s for natural rubber) [30]. It should be noted that the second step of the reaction is relatively slow. This seems to be common feature of all antiozonants. Despite the fact that they are capable of reacting with several ozone molecules, the products of the subsequent reaction steps are ineffective because of the slower rate of the reaction with ozone. The introduction into the system of a variety of anti-ozonants instead of the reference olefin has made it possible to determine, from the depression of SO_2 release, the rate constants of their reactions with ozone. These values together with their rate constants with respect to the ozone dissolution rate are given in Table 8.1.

TABLE 8.1 Rate constants of the reaction between antiozonants and ozone and their protective action in rubber [27].

Compound	Formula	Time to cracking (min)	$k_{21} \times 10^{-6}$ (L/mol s)
N, N'-di-n-jctyl-p-Phenylene-diamine	C_8H_{17}—NH—⟨O⟩—NH—C_8H_{17}	840	7
N, N'-diisopen-tyl-p-pheny-lenediamine	C_5H_{11}—NH—⟨O⟩—NH—C_5H_{11}	870	8
N-Phenyl-N'-iso-propy-p-phenylene-diamine	⟨O⟩—NH—⟨O⟩—NH—$CH(CH_3)_2$	500	7
N, N'-di-α-methylben-zyl-p-phenylene-diamine	⟨O⟩—C(CH₃)(H)—NH—⟨O⟩—NH—C(CH₃)(H)—⟨O⟩	250	5
N-α-meth-yl-benzyl-anisi-dine	⟨O⟩—C(CH₃)(H)—NH—⟨O⟩—OCH_3	90	4
N-butil-N, N'-dibutyl-thio-urea	C_4H_9NH—C(=S)—$N(C_4H_9)$	790	2
Methyl oleate	CH_3O—C(=O)—$(CH_2)_7$—CH=CH—$(CH_2)_7$—CH_3	80	1

The tabulated data indicate that the protective action is exerted only by those compounds which react with ozone at a rate exceeding the rate of the reaction between ozone and the C=C bonds in macromolecules. The greater the reaction rate constant, the more effective the protection by the compounds tested. Figure 8.6 illustrates the relative efficiency τ_{an}/τ_c expressed as the ratio between the time to cracking in stabilized rubber and the time to cracking in the control specimen [31], versus the ratio between the rate constants of the ozone-antiozonant and ozone —>C=C< reactions (k_{an}/k_{12}). It can be seen that the resulting curve is linear in semi-logarithmic coordinates (Eq. 10).

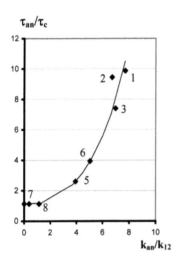 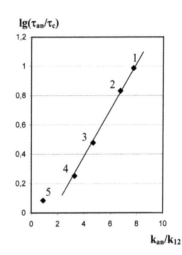

FIGURE 8.6   Relations $\tau_{an}/\tau_c = f(k_{an}/k_{12})$ (a) and $log\tau_{an}/\tau_c = f(k_{an}/k_{12})$ (b). 1, N, N'-dioctyl-p-phenylenediamine; 2, N, N'-di-isopentyl-p-phenylenediamine; 3, N-phenyl-N'-isopropyl-p-phenylenediamine; 4, N, N'-di-α-methylbenzyl-p- phenylenediamine; 5, N-α-methyl-benzylanisidine; 6, N, N, N'-tributylthiourea; 7, 2,2-thio-bis-(6-tret-butil-4-methylphenol); 8, methyl oleate.

$$\log {}^{\tau_{an}}\!/_{\tau_c} = f\left({}^{k_{an}}\!/_{k_{12}}\right) \tag{10}$$

It was established in these experiments that only one class of compounds, namely the thioureas, is excluded from the series. Their experimentally measured activity by this method turned-out to be abnormally high. During tests of tires under normal operating conditions or during dynamic tests this anomaly was not observed. It was natural to assume that the observed deviation is due not the chemical properties of thiourea but to its specific physical behavior in rubbers [18, 22, 32], in particular its tendency to accumulate on the rubber surface to a greater extent than other substances examined. This is also corroborated by published data concentrating the relative content of various antiozonants on the surface when they were initially incorporated in equal amounts [33].

The relation $\log \tau_{an}/\tau_c = f(k_{an}/k_{12})$ is interesting in two respects. Firstly, it can be used as a basis for the development of quantitative methods of evaluating the efficiency of antiozonants. At present no such methods are available; there are only qualitative estimates: good, satisfactory, poor, and so on [32], which make it impossible to compare compounds of different classes or compounds of the same class taken in different amounts. Secondly, it can be an instrument in approximate calculations of maximum antiozonant efficiency. If it is assumed that this relation persists when the rate constant k_{an} increases to a reasonable value (ca 10^8 L/mol s), even rough estimates show that the efficiency of such an antiozonant will exceed that of the currently available ones by several orders of magnitude.

The mechanism of the physical protection of vulcanisates of unsaturated polymers against ozone is much less well understood and theoretical interpretations have not been elaborated. It is known that the introduction of various waxes into the original stock enhances the ozone resistance of vulcanisates, particularly under static conditions [34–36]. Waxes may be used individually and in combination with antiozonants [18, 37], and it is generally considered that waxes form a protective film on the rubber surface, which slows down the reaction between ozone and the C=C bonds of the polymer. The permeability of the film is associated with the temperature of the wax film, the temperature at which it starts to soften, and plasticity [38].

Figure 8.7 shows the results of determining the time to cracking (τ_c) in the case of unfilled natural rubber containing paraffins with different softening points. The efficiency of waxes is higher at low strains; at high strains, the number of cracks decreases in the presence of waxes but then overstrains ahead of crack and crack propagation rate may even increase.

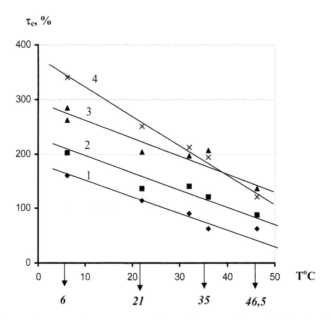

FIGURE 8.7 τ_c (expressed as percentage of τ_c without wax) for unfilled vulcanisation of natural rubber, containing five parts by weight of paraffin wax, at various test temperatures (°C): 1, 25; 2, 35; 3, 45; 4, 57. $\varepsilon = 10\%$; $[O_3] = 4.5 \times 10^{-7}$ mol/L.

The chemical composition of waxes influences the protection efficiency. The presence of C=C bonds, amino groups, metal salts and other additives in waxes increases the time to cracking and time to degradation of the specimen [39].

KEYWORDS

- experimental
- interaction
- kinetics
- mechanism
- ozone
- rubbers

REFERENCES

1. Gorbunov B. N., Gurvitch A.Ya., Maslova I. P. Chimiya Technologiya stabilizatorov Polymernyh materialov. M., Chimiya, 1981.
2. Kuzminskii A. S., Kavun S. M., Kirpichev V. P. Phisiko-chimicheskie osnovy polucheniya, pererabotki I primeneniya elastomerov. M., Chimiya, 1976.
3. Zuev Yu.S., Degteva T. G.., Stoikost elastomerov v ekspluatatcionnyh usloviyah., M., Chimiya, 1986, 264 p.
4. Franco Cataldo. Polymer Degradation and Stability, 2001, v.73, p.511–520.
5. Muhutdinov E. R., Diakonov G. S. Vestnik Kazanskogo Universiteta, 2010, N10, pp 483–504.
6. Ionova N. I., Zemskii D. N., Dorofeeva Yu.N., KurliS. K., Mohnatkina E. G. Kautchuk i Rezina., 2011, 1, 9–12.
7. Razumovskii S. D., Rakovskii S. K., Shopov D. M., Zaikov G. E. Ozone and its reactions with organic compounds, Bulgarian Ak.Nauk, Sofia, 1983, 289 p.
8. Veith, A. G., *Rubber Chem. Technol.*, 45 (1972) 293.
9. Lipkin, A. M., Grinberg, A. E., Gurvich, Ya. A., Zolotarevskaya, L. K., Razumovskii, S. D. and Zaikov G. E., *Vysokomol. Soedin.*, A14 (1972) 78.
10. Lake, G. I., *Rubber Chem. Technol.*, 43 (1970) 1230.
11. Zuev, Yu. S. and Pravednikov, S. I., *Kauch. Rezina*, 1 (1961) 30.
12. Razumovskii, S. D. and Batashova, L. S., *Vysokomol.Soedin.*, A9 (1969) 588.
13. Zuev, Yu. S., Koshelev, F. F., Otopkova, M. A. and Mikhaleva, S. B., *Kauch. Rezina*, 8 (1965) 12.
14. Lorenz, O and Parks, C. R., *Rubber Chem. Technol.*, 36 (1963) 194.
15. Layer, R. W., *Rubber Chem. Technol.*, 39, (1966) 1584.
16. Razumovskii, S. D., Buchachenko, A. L., Shapiro, A. B., Rozantsev, E. G. and Zaikov, G. E., *Dokl. Akad. Nauk SSSR*, 183 (1968) 1106.
17. Murray, R. W. and Story, P. R. in Chemical Reactions of Polymers, Vol. 2, Ed. E. M. Fettes, (1964), Wiley-Interscience, New York.
18. Zuev, Yu. S., *Razrushenie Polymerov pod Deistviem Agressivnykh sre,* (1972). Khimiya, Moscow. (a) Ibid. p. 103.
19. Lipkin, A. M., Zolotarevskaya, L., K., Grinberg, A. E., Gurvich, Ya. A., Razumovskii, S. D. and Zaikov, G. E., *Vysokomol.Soedin.*, A14 (1972) 680.
20. Ambelang J. C., Kline, R. H., Lorenz, O. M. and Parks, C. R., *Rubber Chem.Technol.*, 36 (1963) 1497.
21. Tucker, H. *Rubber Chem.Technol.*, 32 (1959) 269.
22. Zuev, Yu. S., *Zhurn. Vses. Khim. Obschchestvo Mendeleeva*, 11 (1965) 288.
23. *Ozone. Chemistry and Technology.* A Review of the Literature, 1961–1974 (1976). Philadelphia.
24. Braden, M., *J. Appl. Polym. Sci.,* 6 (1962) 86.
25. Delman, A. D., Sims, B. B. and Allison, A. R., *J. Anal. Chem.*, 26 (1954) 1589.
26. Razumovskii, S. D. and Zaikov, G. E., *Ozone and its Reactions with Organic Compounds* (1974), Nauka, Moscow.
27. Lipkin, A. M., Razumovskii, S. D. Gurvich, Ya. A., Grinberg, A. E., and Zaikov, G. E., *Dokl. Akad. Nauk SSSR*, 192 (1970) 127.

28. Parfenov, V. M., Rakovski, S. K., Shopov, D., M., Popov, A. A. and Zaikov, G. E., *Izves. Khim. Bolg. Akad. Nauk,* 11 (1978) 180.
29. Rakovski, S. K., Cherneva, D. R., Shopov, D. M. and Razumovskii, S. D., *Izves. Khim. Bolg. Akad. Nauk,* 9 (1976) 711.
30. Kefeli, A. A., Vinitskaya, E. A., Markin, V. S., Razumovskii, S. D., Gurvich, Ya.A., Lipkin, A. M. and Neverov, A. N., *Vysokomol. Soedin.,* A19 (1977) 2633.
31. Lipkin, A. M., Grinberg, A. E., Gurvich, Ya.A., Zolotarevskaya, L. K., Razumovskii, S. D., Zaikov, G. E., Vysokomolek.soedin., A14, (1972) 87.
32. Ambelang, I. C. and Lorenz, O., *Rubber Chem. Technol.,* 36–91963) 1533.
33. Hogkinson, G. T. and Kendall, C. E., in Proc. 5th Rubber Technol.Conf., Ed. T. H. Messenger, (1962). Institution of the Rubber Industry, London.
34. Bennet, H., Commercial Waxs, 2nd edn. (19560. Chemical Publishing Co., New York.
35. Backer, D. E., *Rubber Age (NY),* 77 (1955) 58.
36. Buswell, A. G., Watts, J. T., *Rubber Chem. Technol.,* 35 (1962) 421.
37. Zinchenko, N. P., Vinogradova, T. N., *Zashchita Shinnykh Rezin ot Vozdeistviya Ozona I Utomleniya* (1969). TsNIIEneftekhim, Moscow.
38. Zuev, Yu. S., Karandashev, B. P., *Plastifikatory I Zashchitnye Agenty iz Neftyanogo Syrya,* Ed.I. P. Lukashevich, (1970) p.161. Khimiya, Moscow.
39. Zuev, Yu. S., Postovskaya, A. F., *ibid,* p.136.

CHAPTER 9

THE INTERCOMMUNICATION OF FRACTAL ANALYSIS AND POLYMERIC CLUSTER MEDIUM MODEL

G. V. KOZLOV, I. V. DOLBIN, JOZEF RICHERT, O. V. STOYANOV, and G. E. ZAIKOV

CONTENTS

9.1   INTRODUCTION

This chapter reviews about the intercommunication of fractal analysis and polymeric cluster medium model. It is shown that close intercommunication exists between notions of local order and fractality in a glassy polymers case, having key physical grounds and expressed by the simple analytical relationship. It has also been shown, that the combined usage of these complementing one another concepts allows to broaden possibilities of polymeric mediums structure and properties analytical description. The indicated expressions will be applied repeatedly in further interpretation.

9.2   BACKGROUND

During the last 25 years a fractal analysis methods obtained wide spreading in both theoretical physics [1] and material science [2], in particular, in physics-chemistry of polymers [3–8]. This tendency can be explained by fractal objects wide spreading in nature.

There are two main physical reasons, which define intercommunication of fractal essence and local order for solid-phase polymers: the thermodynamically nonequilibrium and dimensional periodicity of their structure. In Ref. [9] the simple relationship was obtained between thermodynamically nonequilibrium characteristic – Gibbs function change at self-assembly (cluster structure formation of polymers $\Delta \tilde{G}^{im}$ – and clusters relative fraction φ_{cl} in the form:

$$\Delta \tilde{G}^{im} \sim \phi_{cl}.\tag{1}$$

This relationship graphic interpretation for amorphous glassy polymers – polycarbonate (PC) and polyarylate (PAr) – is adduced in Fig. 9.1. Since at $T = T_g (T_m)$ (where T, T_g and T_m are testing, glass transition and melting temperatures, accordingly) $\Delta \tilde{G}^{im} = 0$ [10, 11], then from the relationship (1.1) it follows, that at the indicated temperatures cluster structure full decay ($\varphi_{cl} = 0$) should be occurred or transition to thermodynamically equilibrium structure.

As for the intercommunication of parameters, characterizing structure fractality and medium thermodynamically nonequilibrium, it should exist indisputably, since precisely nonequilibium processes formed fractal structures. Solid bodies fracture surfaces analysis gives evidence of such rule fulfillment – a large number of experimental papers shows their fractal structure, irrespective of the analyzed material thermodynamically state [12]. Such phenomenon is due to the fact, that the fracture process is thermodynamically nonequilibrium one [13]. Polymers structure fractality is due to the same circumstance. The experimental confirmation can be found in Refs. [14–16]. As for each real (physical) fractal, polymers structure fractal properties are limited by the defined linear scales. So, in Refs. [14, 17] these scales were determined within the range of several Ångströms (from below) up to several tens Ångströms (from above). The lower limit is connected with medium structural elements finite size and the upper one – with structure fractal dimension d_f limiting values [18]. The indicated above scale limits correspond well to cluster nanostructure specific boundary sizes: the lower—with statistical segment length l_{st}, the upper—with distance between clusters R_{cl} [19].

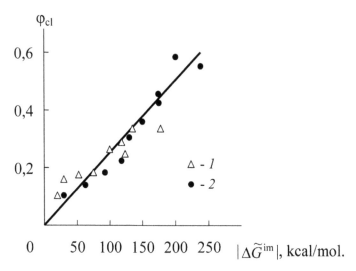

FIGURE 9.1 The dependence of clusters relative fraction φ_{cl} on absolute value of specific Gibbs function of nonequilibrium phase transition $\Delta \tilde{G}^{im} \|$ for amorphous glassy polymers—polycarbonate (1) and polyarylate (2) [3].

Polymeric medium's structure fractality within the indicated above scale limits assumes the dependence of their density ρ on dimensional parameter L (see Fig. 9.2) as follows [1]:

$$\rho \sim L^{d_f - d},\tag{2}$$

where d is dimension of Euclidean space, into which a fractal is introduced.

In Fig. 9.3, amorphous polymers nanostructure cluster model is presented. As one can see, within the limits of the indicated above dimensional periodicity scales Figs. 9.2 and 9.3 correspond each other, that is, the cluster model assumes ρ reduction as far as possible from the cluster center. Let us note that well-known Flory "felt" model [20] does not satisfy this criterion, since for it $\rho \approx$ const. Since, as it was noted above, polymeric mediums structure fractality was confirmed experimentally repeatedly [14–16], then it is obvious, that cluster model reflects real solid-phase polymers structure quite plausibly, whereas "felt" model is far from reality. It is also obvious, that opposite intercommunication is true—for density ρ finite values change of the latter within the definite limits means obligatory availability of structure periodicity.

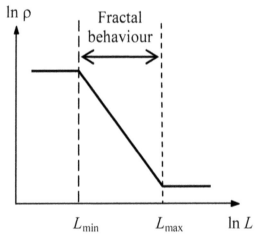

FIGURE 9.2   The schematic dependence of density ρ on structure linear scale L for solid body in logarithmic coordinates. The fractal structures are formed within the range L_{min}–L_{max} [18].

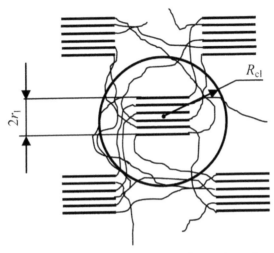

FIGURE 9.3   The model of cluster structure for amorphous polymeric media [7].

Meanwhile, we must not over look the fact that fractal analysis gives only polymers structure general mathematical description only, that is, does not identify those concrete structural units (elements), which any real polymer consists. At the same time, the physical description of polymers thermodynamically nonequilibrium structure within the frameworks of local order notions is given by the cluster model of amorphous polymers structure gives, which is capable of its elements quantitative identification [21].

The well described in literature experimental studies was served as premise for the cluster model development. As it is known (see, e.g., [22, 23]), high-molecular amorphous polymeric mediums, which are in glassy state, at their deformation above glass transition temperature can display high-elastic behavior (when high-elasticity plateau is formed), which can be described by the high-elasticity theory rules. Personally, Langievene [24, 25] and Gauss [26] equations can be used for polymers behavior description large strains. In the latter it is assumed that at deformation in plateau region polymeric chain is not stretched fully and the relationship between true stress σ^{tr} and drawing ratio λ at uniaxial tension is written as follows:

$$str = Gp(L2 - L-1), \tag{3}$$

where G_p is the so-called strain hardening modulus.

The value G_p knowledge allows formally to calculate the density v_e of macromolecular entanglements network in polymeric medium according to the well-known expressions of high-elasticity theory [23]:

$$M_e = \frac{\rho RT}{G_p}, \tag{4}$$

$$v_e = \frac{\rho N_A}{M_e}, \tag{5}$$

where ρ is polymer density, R is universal gas constant, T is testing temperature, M_e is molecular weight of chain part between entanglements, N_A is Avogadro number.

However, the attempts to estimate the value M_e (or v_e) according to the determined from Eq. (3) G_p values result to unlikely low calculated values M_e (or unreal high values v_e), contradictoring to Gaussian statistics requirements, which assume availability on chain part between entanglements no less than ~ 13 monomer links [27].

As alternative the authors [28] assumed, that besides the indicated above binary hooking network in polymers glassy state another entanglements type was available, nodes of which by their structure were similar to crystallites with stretched chains (CSC). Such entanglements node possesses large enough functionality F (under node functionality emerging from it chains number is assumed [29] and it was called cluster. Cluster consists of different macromolecules segments and each such segment length is postulated equal to statistical segment length l_{st} ("stiffness part" of chain [30]). In that case the effective (real) molecular weight of chain part between clusters M_{cl}^{ef} can be calculated as follows [29]:

$$M_{cl}^{ef} = \frac{M_{cl}F}{2}, \tag{6}$$

where M_{cl} is molecular weight of the chain part between clusters, calculated according to the Eq. (4).

It is obvious, that at large enough F the reasonable values M_{cl}^{ef}, satisfied to Gaussian statistics requirements, can be obtained. Further on for parameters of entanglements cluster network and macromolecular binary hookings network distinction we will used the indices "cl" and "e", accordingly. Therefore, the offered in paper [28] model assumes, that amorphous polymer structure represents itself local order regions (domains), consisting of different macromolecules collinear densely packed segments (clusters), immersed in loosely packed matrix. Simultaneously clusters play the role of physical entanglements network multifunctional nodes. The value F (within the frameworks of high-elasticity theory) can be estimated as follows [31]:

$$F = \frac{2G_\infty}{kTv_{cl}} + 2 ,$$

(7)

where G_∞ is equilibrium shear modulus, k is Boltzmann constant, v_{cl} is cluster network density.

In Fig. 9.4, the dependences $v_{cl}(T)$ for polycarbonate (PC) and polyarylate (PAr) are adduced. These dependences show v_{cl} reduction at T growth, that assumes local order regions (clusters) thermofluctuational nature. Besides, on the indicated dependences two characteristic temperatures are found easily. The first from them, glass transition temperature T_g, defines clusters full decay (see also, Fig. 9.1), the second T_g', corresponds to the fold on curves $v_{cl}(T)$ and settles down on about 50 K lower T_g.

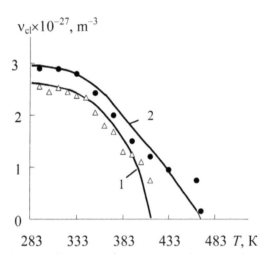

FIGURE 9.4   The dependences of macromolecular entanglements cluster network density ν_{cl} on testing temperature T for PC(1) and PAr (2) [28]

Earlier within the frameworks of local order concepts it has been shown that temperature T_g is associated with segmental mobility releasing in polymer loosely packed regions. This means, that within the frameworks of cluster model T_g can be associated with loosely packed matrix devitrification. The dependences $F(T)$ for the same polymers have a similar form (Fig. 9.5).

Two main models, describing local order structure in polymers (with folded [34] and stretched [28] chains) allows to make common conclusion: local order regions in polymer matrices play role of macromolecular entanglements physical network nodes [28, 34]. However, their reaction on mechanical deformation should be distinguished essentially, it at large strains regions with folded chains ("bundles") are capable to unfolding of folds in straightened conformations, then clusters have no such possibility and polymer matrix deformation can occur by "tie" chains (connecting only clusters) straightening, that is, by their orientation in the applied stress direction.

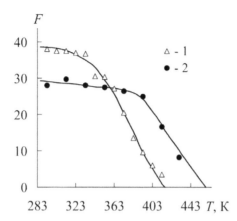

FIGURE 9.5   The dependence of clusters functionality F on testing temperature T for PC(1) and PAr (2) [28].

Turning to analogy about mediums with crystalline morphology of polymer matrices, let us note, that semi-crystalline polymers (e.g., high-molecular polyethylenes) large strains, making 1000 2000%, are realized owing to crystallites folds unfolding [35].

Proceeding from the said above and analyzing values of polymers limiting strains, one can obtain the information about local order regions type in amorphous and semi-crystalline polymers. The fulfilled by the authors [36] calculations have show that the most probable type of local nano-structures in amorphous polymer matrix is an analog of crystallite with stretched chains, that is, cluster.

Let us note, completing this topic, that for the local order availability substantiation in amorphous polymer matrix (irrespective to concrete structural model of medium) strict mathematical proofs of the most common character exist. For example, according to the proved in numbers theory Ramsey's theorem any large enough quantity $N_i > R(i, j)$ of numbers, points or objects (in the considered case—statistical segments) contains without fail high-ordered system from $N_j \leq R(i, j)$ such segments. Therefore absolute disordering of large systems (structures) is impossible [37, 48].

As it is known [39], structures, which behave themselves as fractal ones on small length scales and as homogeneous ones—on large ones, are named homogeneous fractals. Percolation clusters near percolation thresh-old are such fractals [1]. As it will be shown lower, cluster structure is a percolation system and in virtue of the said above—homogeneous fractal. In other words, local order availability in polymers condensed state testi-fies to there structure fractality [21].

The percolation system fractal dimension d_f can be expressed as fol-lows [39]:

$$d_f = d - \frac{\beta_p}{v_p},$$

$$\text{(8)}$$

where β_p and v_p are critical exponents (indices) in percolation theory.

Hence, the condition $\beta_p \neq 0$, which follows from the data of Table 9.1, also testifies to also polymer matrix structure fractality. It is obvious, that polymer medium structure fractality can be determined by the dependence of φ_{cl} on temperature, that is, using analysis of local order thermofluctua-tional effect (see Fig. 9.4). Let us note that structure fractality and "freez-ing" below T_g local order (φ_{cl} = const) from the physical point of view are intercepting notions.

TABLE 9.1   A percolation clusters characteristics [5].

Parameter	Experimentally determined magni-tudes		Theoretical "geometrical" magnitudes
	EP-1[a]	EP-2[a]	
β_p	0.54	0.58	0.40
v	1.20	1.15	0.88
β_p/v_p	0.45	0.50	0.46
d_f	2.55	2.50	2.55

[a]EP-1 and EP-2 are epoxy polymers on the basis of resin ED-22.

The considered above principles allow to connect by analytical rela-tionship the fractal dimension d_f and local order characteristics φ_{cl} (or v_{cl}). According to the Ref. [18] the value d_f is connected with Poisson's ratio v as follows:

$$df = (d-1)(1+n), \tag{9}$$

and the value v is connected, in its turn, with v_{cl} according to the relationship [21]:

$$v = 0,5 - 3,22 \times 10^{-10} \left(l_0 v_{cl} \right)^{1/2}, \tag{10}$$

where the following intercommunication between φ_{cl} exists:

$$\varphi_{cl} = Sl_0 C_\infty \varphi_{cl}, \tag{11}$$

where S is macromolecule cross-sectional area, l_0 is main chain skeletal length, C_∞ is characteristic ratio.

From the expressions (9) ÷ (11) the dependence of d_f on φ_{cl} or v_{cl} can be obtained for the most often used case $d = 3$ [21]:

$$d_f = 3 - 6,44 \times 10^{-10} \left(\frac{\phi_{cl}}{SC_\infty} \right)^{1/2} \tag{12}$$

$$\text{or } d_f = 3 - 6,44 \times 10^{-10} \left(l_0 v_{cl} \right)^{1/2}. \tag{13}$$

Let us note, returning to the Eq. (1), that its substation in Eq. (12) allows to obtain the following relationship:

$$\Delta \tilde{G}^{im} \sim C_\infty S \left(3 - d_f \right)^2. \tag{14}$$

The value $\Delta \tilde{G}^{im}$ (Gibbs specific function of local, for example, supra-molecular structures formation) is given for nonequilibrium phase transition "supercooled liquid ® solid body" [11]. From the Eq. (14) it follows, that the condition $\Delta \tilde{G}^{im} = 0$ is achieved at $d_f = 3$, that is, at $d_f = d$ and at transition to Euclidean behavior. In other words, a fractal structures are formed only in nonequilibrium processes course that is noted earlier [12].

Above general reasoning's about possibility of existence in polymeric media of local order regions, based on the Ramsey theorem, were presented in a simplified enough form. It can be shown similarly, that any structure, consisting of N elements, at $N > B_N(j)$ represents it self totality of finite number $k \leq j$ of put into each other self-similar structures, Hausdorff dimension of which in the general case can be different one. Therefore, any structure irrespective of its physical nature, consisting of a large enough elements number, can be represented as multifractal (in partial case as monofractal) and described by spectrum of Renyi dimensions d_q, $q = -\infty \div +\infty$ [18]. In paper [40] it has been shown, that the condensed systems attainment to self-organization in scale-invariant multifractal forms is the result of key principles of open systems thermodynamics and d_q is defined by competition of short- and long-range interatomic correlations, determining volume compressibility and shear stiffness of solid bodies, accordingly.

9.3   CONCLUSION

It can be noted as a brief resume to this chapter, that close intercommunication exists between notions of local order and fractality in a glassy polymers case, having key physical grounds and expressed by the simple analytical relationship (the Eqs. (12) and (13)). It has also been shown, that the combined usage of these complementing one another concepts allows to broaden possibilities of polymeric mediums structure and properties analytical description. The indicated expressions will be applied repeatedly in further interpretation.

KEYWORDS

- fractal analysis
- glassy polymer
- intercommunication
- medium model
- polymeric cluster
- theoretical

REFERENCES

1. Feder, E. Fractals. New York, Plenum Press, 1989, 248 p.
2. Ivanova, V. S., Balankin, A. S., Bunin, I.Zh., Oksogoev, A. A. Synergetics and Fractals in Material Science. Moscow, Nauka, 1994, 383 p.
3. Novikov, V. U., Kozlov, G. V. A Macromolecules Fractal Analysis. Uspekhi Khimii, 2000, v. 69, № 4, p. 378–399.
4. Novikov, V. U., Kozlov, G. V. Structure and Properties of Polymers within the Frameworks of Fractal Approach. Uspekhi Khimii, 2000, v. 69, № 6, p. 572–599.
5. Kozlov, G. V., Novikov, V. U. Synergetics and Fractal Analysis of Cross-Linked Polymers. Moscow, Klassika, 1998, 112 p.
6. Kozlov, G. V., Yanovskii Yu.G., Zaikov, G. E. Synergetics and Fractal Analysis of Polymer Composites Filled with short Fibers. New York, Nova Science Publishers, Inc, 2011, 223 p.
7. Kozlov, G. V., Yanovskii Yu.G., Zaikov, G. E. Structure and Properties of Particulate-Filled Polymer composites: The Fractal Analysis. New York, Nova Science Publishers, Inc, 2010, 282 p.
8. Shogenov, V. N., Kozlov, G. V. A Fractal Clusters in Physics-Chemistry of Polymers. Nal'chik, Poligrafservis I T, 2002, 268 p.
9. Kozlov, G. V., Zaikov, G. E. The Generalized Description of Local Order in Polymers. In book: Fractal and Local order in Polymeric Materials. Ed. Kozlov, G. V., Zaikov, G. E. New York, Nova Science Publishers, Inc, 2001, p. 55–64.
10. Gladyshev, G. P., Gladyshev, D. P. About Physical-Chemical Theory of Biological Evolution (Preprint). Moscow, Olimp, 1993, 24 p.
11. Gladyshev, G. P., Gladyshev, D. P. The Approximative Thermodynamical Equation for Nonequilibrium Phase Transitions. Zhurnal Fizicheskoi Khimii, 1994, v. 68, № 5, p. 790–792.
12. Hornbogen, E. Fractals in Microstructure of Metals. Int. Mater. Rev, 1989, v. 34, № 6, p. 277–296.
13. Bessendorf, M. H. Stochastic and Fractal Analysis of Fracture Trajectories. Int. J. Engng. Sci., 1987, v. 25, № 6, p. 667–672.
14. Zemlyanov, M. G., Malinovskii, V. K., Novikov, V. N., Parshin, P. P., Sokolov, A. P. The Study of Fractions in Polymers. Zhurnal Eksperiment. I Teoretich. Fiziki, 1992, v. 101, № 1, p. 284–293.
15. Kozlov, G. V., Ozden, S., Krysov, V. A., Zaikov, G. E. The Experimental Determination of a Fractal Dimension of the Structure of Amorphous Glassy Polymers. In book: Fractals and Local Order in Polymeric Material. Ed. Kozlov, G. V., Zaikov, G. E. New York, Nova Science Publishers, Inc., 2001, p. 83–88.
16. Kozlov, G. V., Ozden, S., Dolbin, I. V. Small Angle X-Ray Studies of the Amorphous Polymers Fractal Structure. Russian Polymer News., 2002, v. 7, № 2, P. 35–38.
17. Bagryanskii, V. A., Malinovskii, V. K., Novikov, V. N., Pushchaeva, L. M., Sokolov, A. P. Inelastic Light Diffusion on Fractal Oscillation Modes in Polymers. Fizika Tverdogo Tela, 1988, v. 30, № 8, p. 2360–2366.
18. Balankin, A. S., Synergetics of Deformable Body. Moscow, Publishers of Defence Ministry of SSSR, 1991, 404p.

19. Kozlov, G. V., Belousov, V. N., Mikitaev, A. K. Description of Solid Polymers as Quasitwophase Bodies. Fizika I Tekhnika Vysokikh Davlenii, 1998, v. 8, № 1, p. 101–107.
20. Flory, P. J. Spatial Configuration of Macromolecular Chains. Brit. Polymer, J., 1976, v. 8, № 1, p. 1–10.
21. Kozlov, G. V., Ovcharenko, E. N., Mikitaev, A. K. Structure of the Polymers Amorphous State. Moscow, Publishers of the, D. I.Mendeleev RkhTU, 2009, 392 p.
22. Haward, R. N. The Application of a Gauss-Eyring Model to Predict the Behavior of Thermoplastics in Tensile Experiment. J. Polymer Sci.: Part B: Polymer Phys, 1995, v. 33, № 8, p. 1481–1494.
23. Haward, R. N. The Application of a Simplified Model for the Stress-Strain curves of Polymers. Polymer, 1987, v. 28, № 8, p. 1485–1488.
24. Boyce, M. C., Parks, D. M., Argon, A. S. Large Inelastic Deformation in Glassy Polymers. Part, I. Rate Dependent Constitutive Model. Mech. Mater, 1988, v. 7, № 1, p. 15–33.
25. Boyce, M. C., Parks, D. M., Argon, A. S. Large Inelastic Deformation in Glassy Polymers. Part II. Numerical Simulation of Hydrostatic Extrusion. Mech. Mater, 1988, v. 7, № 1, p. 35–47.
26. Haward, R. N. Strain Hardening of Thermoplastics. Macromolecules, 1993, v. 26, № 22, p. 5860–5869.
27. Bartenev, G. M., Frenkel, S.Ya. Physics of Polymers. Leningrad, Khimiya, 1990, 432 p.
28. Belousov, V. N., Kozlov, G. V., Mikitaev, A. K., Lipatov Yu.S. Entanglements in Glassy State of Linear Amorphous Polymers. Doklady AN SSSR, 1990, v. 313, № 3. p. 630–633.
29. Flory, P. J. Molecular Theory of Rubber Elasticity. Polymer, J., 1985, v. 17, № 1, p. 1–12.
30. Bernstein, V. A., Egorov, V. M. Differential Scanning Calorimetry in Physics-Chemistry of the Polymers. Leningrad, Khimiya, 1990, 256 p.
31. Graessley, W. W. Linear Viscoelasticity in Gaussian Networks. Macromolecules, 1980, v. 13, № 2, p. 372–376.
32. Perepechko, I. I., Startsev, O. V. Multiplet Temperature Transitions in Amorphous Polymers in Main Relaxation Region. Vysokomolek. Soed. B, 1973, v. 15, № 5, p. 321–322.
33. Belousov, V. N., Kotsev, B.Kh., Mikitaev, A. K. Two-step of Amorphous Polymers Glass Transition Doklady. AN SSSR, 1983, v. 270, № 5, p. 1145–1147.
34. Arzhakov, S. A., Bakeev, N. F., Kabanov, V. A. Supramolecular Structure of Amorphous Polymers. Vysokomolek. Soed. A, 1973, v. 15, № 5, p. 1145–1147.
35. Marisawa, Y. The Strength of Polymeric Materials. Moscow, Khimya, 1987, 400 p.
36. Kozlov, G. V., Sanditov, D. S., Serdyuk, V. D. About Suprasegmental Formations in Polymers Amorphous State. Vysokomolek. Soed. B, 1993, v. 35, № 12, p. 2067–2069.
37. Mikitaev, A. K., Kozlov, G. V. The Fractal Mechanics of Polymeric Materials. Nal'chik, Publishers KBSU, 2008, 312 p.
38. Balankin, A. S., Bugrimov, A. L., Kozlov, G. V., Mikitaev, A. K., Sanditov, D. S. The Fractal Structure and Physical Mechanical Properties of Amorphous Glassy Polymers. Doklady AN, 1992, v. 326, № 3, p. 463–466.

39. Sokolov, I. M. Dimensions and other Geometrical Critical Exponents in Percolation Theory. Uspekhi Fizicheshikh Nauk, 1986, v. 150, № 2, p. 221–256.
40. Balankin, A. S. Fractal Dynamics of Deformable Mediums. Pis ma v ZhTF, 1991, v. 17, № 6. p. 84–89.

CHAPTER 10

POLYMERS AS NATURAL COMPOSITES: STRUCTURE AND PROPERTIES

G. V. KOZLOV, I. V. DOLBIN, JOZEF RICHERT, O. V. STOYANOV, and G. E. ZAIKOV

CONTENTS

10.1   INTRODUCTION

The stated in the present article results give purely practical aspect of such theoretical concepts as the cluster model of polymers amorphous state stricture and fractal analysis application for the description of structure and properties of polymers, treated as natural nanocomposites. The necessary nanostructure goal-directed making will allow to obtain polymers, not yielding (and even exceeding) by their properties to the composites, produced on their basis. Structureless (defect-free) polymers are imagined the most perspective in this respect. Such polymers can be natural replacement for a large number of elaborated at present polymer nanocomposites. The application of structureless polymers as artificial nanocomposites polymer matrix can give much larger effect. Such approach allows to obtain polymeric materials, comparable by their characteristics with metals (e.g., with aluminum).

The idea of different classes polymers representation as composites is not new. Even 35 years ago Kardos and Raisoni [1] offered to use composite models for the description of semicrystalline polymers properties number and obtained prediction of the indicated polymers stiffness and thermal strains to a precision of ±20%. They considered semicrystalline polymer as composite, in which matrix is the amorphous and the crystallites are a filler. The authors [1] also supposed that other polymers, for example, hybrid polymer systems, in which two components with different mechanical properties were present obviously, can be simulated by a similar method.

In Ref. [2] it has been pointed out, that the most important consequence from works by supramolecular formation study is the conclusion that physical-mechanical properties depend in the first place on molecular structure, but are realized through supramolecular formations. At scales interval and studies methods resolving ability of polymers structure the nanoparticle size can be changed within the limits of $1 \div 100$ and more nanometers. The polymer crystallites size makes up $10 \div 20$ nm. The macromolecule can be included in several crystallites, since at molecular weight of order of 6×10^4 its length makes up more than 400 nm. These reasoning's point out, that macromolecular formations and polymer systems in virtue of their structure features are always nanostructural systems.

However, in the cited above works the amorphous glassy polymers consideration as natural composites (nanocomposites) is absent, although they are one of the most important classes of polymeric materials. This gap reason is quite enough, that is, polymers amorphous state quantitative model absence. However, such model appearance lately [3–5] allows to consider the amorphous glassy polymers (both linear and cross-linked ones) as natural nanocomposites, in which local order regions (clusters) are nanofiller and surrounded them loosely packed matrix of amorphous polymers structure is matrix of nanocomposite. Proceeding from the said above, in the present chapter description of amorphous glassy polymers as natural nanocomposites, their limiting characteristics determination and practical recommendation by the indicated polymers properties improvement will be given.

10.2   NATURAL NANOCOMPOSITES STRUCTURE

The synergetics principles revealed structure adaptation mechanism to external influence and are universal ones for self-organization laws of spatial structures in dynamical systems of different nature. The structure adaptation is the reformation process of structure, which loses stability, with the new more stable structure self-organization. The fractal (multifractal) structure, which is impossible to describe within the framework of Euclidean geometry, are formed in reformation process. A wide spectrum of natural and artificial topological forms, the feature of which is self-similar hierarchically organized structure, which amorphous glassy polymers possessed [6], belongs to fractal structures.

The authors of Refs. [7, 8] considered the typical amorphous glassy polymer (polycarbonate) structure change within the frameworks of solid body synergetics.

The local order region, consisting of several densely packed collinear segments of various polymer chains (for more details *see* Chapter 9) according to a signs number should be attributed to the nanoparticles (nanoclusters) [9]:

1) their size makes up $2 \div 5$ nm;

2) they are formed by self-assemble method and adapted to the external influence (e.g., temperature change results to segments number per one nanocluster change);

3) the each statistical segment represents an atoms group and boundaries between these groups are coherent owing to collinear arrangement of one segment relative to another.

The main structure parameter of cluster model-nanoclusters relative fraction φ_{cl}, which is polymers structure order parameter in strict physical sense of this tern, can be calculated according to the equation (*see* Chapter 9). In its turn, the polymer structure fractal dimension d_f value is determined according to the equations (*see* Chapter 9).

In Fig. 10.1, the dependence of φ_{cl} on testing temperature T for PC is shown, which can be approximated by the broken line, where points of folding (bifurcation points) correspond to energy dissipation mechanism change, coupling with the threshold values φ_{cl} reaching. So, in Fig. 10.1, T_1 corresponds to structure "freezing" temperature T_0 [4], T_2 to loosely packed matrix glass transition temperature T_g [11] and T_3 to polymer glass transition temperature T_g.

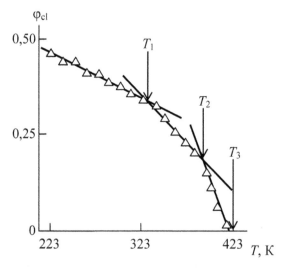

FIGURE 10.1 The dependence of nanoclusters relative fraction φ_{cl} on testing temperature T for PC. The critical temperatures of bifurcation points are indicated by arrows (explanations are given in the text) [18].

Within the frameworks of solid body synergetics it has been shown [12] that at structures self-organization the adaptation universal algorithm [12] is realized at transition from previous point of structure instability to subsequent one. The value $m = 1$ corresponds to structure minimum adaptively and $m = m*$ to maximum one. In Ref. [12] the table is adduced, in which values A_m, m and Δ_i are given, determined by the gold proportion rule and corresponding to spectrum of structure stability measure invariant magnitudes for the alive and lifeness nature systems. The indicated table usage facilitates determination of the interconnected by the power law stability and adaptively of structure to external influence [12].

Using as the critical magnitudes of governing parameter the values φ_{cl} in the indicated bifurcation points T_0, T_g' and T_g (ϕ'_{cl} and T^*_{cl}, accordingly) together with the mentioned above table data [12], values A_m, Δ_i and for PC can be obtained, which are adduced in Table 10.1. As it follows from the data of this table, systematic reduction of parameters A_m and Δ_i at the condition $m = 1 = $ const is observed. Hence, within the frameworks of solid body synergetics temperature T_g' can be characterized as bifurcation point ordering-degradation of nanostructure and T_g as nanostructure degradation-chaos [12].

It is easy to see, that Δ_i decrease corresponds to bifurcation point critical temperature increase.

TABLE 10.1   The critical parameters of nanocluster structure state for PC [8].

The temperature range	ϕ'_{cl}	ϕ^*_{cl}	A_m	Δ_i	m	m^*
213÷333 K	0.528	0.330	0.623	0.618	1	1
333÷390 K	.330	0.153	0.465	0.465	1	2
390÷425 K	0.153	0.049	0.324	0.324	1	8

Therefore, critical temperatures T_{cr} (T_0, T_g' and T_g) values increase should be expected at nanocluster structure stability measure Δ_i reduction. In Fig. 10.2 the dependence of T_{cr} in Δ_i reciprocal value for PC is adduced, on which corresponding values for polyarylate (PAr) are also plotted. This correlation proved to be linear one and has two characteristic points. At Δ_i = 1 the linear dependence $T_{cr}(\Delta_i^{-1})$ extrapolates to $T_{cr} = 293K$, that is, this means, that at the indicated Δ_i value glassy polymer turns into rubber-like state at the used testing temperature $T = 293$K. From the data of the determined by gold proportion law $\Delta_i = 0.213$ at $m = 1$ follows [12]. In the plot of Fig. 10.2, the greatest for polymers critical temperature $T_{cr} = T_{ll}$. (T_{ll} is the temperature of "liquid 1 to liquid 2" transition), defining the transition to "structureless liquid" [13], corresponds to this minimum Δmagnitude. For polymers this means the absence of even dynamical short-lived local order [13].

Hence, the stated above results allow to give the following interpretation of critical temperatures T_g' and T_g of amorphous glassy polymers structure within the frameworks of solid body synergetics. These temperatures correspond to governing parameter (nanocluster contents) φ_{cl} critical values, at which reaching one of the main principles of synergetics is realized-subordination principle, when a variables set is controlled by one (or several) variable, which is an order parameter. Let us also note reformations number $m = 1$ corresponds to structure formation mechanism particle-cluster [4, 5].

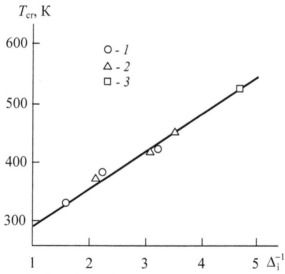

FIGURE 10.2   The dependence of critical temperatures T_{cr} on reciprocal value of nanocluster structure stability measure Δ_i for (1) PC and (2) PAr, (3) T_{ll} value for PC [19].

The authors of Refs. [14, 15] considered synergetics principles application for the description of behavior of separate nanocluster structure, characterized by the integral parameter φ_{cl} nanoclusters in the system for the same amorphous glassy polymers. This aspect is very important, since, as it will be shown is subsequent sections, just separate nanoclusters characteristics define natural nanocomposites properties by critical mode. One from the criterions of nanoparticle definition has been obtained in paper [16]: atoms number N_{at} in it should not exceed 10^3 10^4. In Ref. [15] this criterion was applied to PC local order regions, having the greatest number of statistical segments $n_{cl} = 20$. Since nanocluster is amorphous analog of crystallite with the stretched chains and at its functionality F a number of chains emerging from it is accepted, then the value n_{cl} is determined as follows [4]:

$$n_{cl} = \frac{F}{2},$$

(1)

where the value F was calculated according to the Eq. (7) in Chapter 1.

The statistical segment volume simulated as a cylinder, is equal to $l_{st}S$ and further the volume per one atom of substance (PC) a^3 can be calculated according to the following equation [17]:

$$a^3 = \frac{M}{\rho N_A p},$$

(2)

where M is repeated link molar mass, ρ is polymer density, N_A is Avogadro number, p is atoms number in a repeated link.

For PC $M = 264$ g/mole, $\rho = 1200$ kg/m^3 and $p = 37$. Then $a^3 = 9{,}54$ Å^3 and the value N_{at} can be estimated according to the following simple equation [17]:

$$N_{at} = \frac{l_{st} \cdot S \cdot n_{cl}}{a^3}.$$

(3)

For PC $N_{at} = 193$ atoms per one nanocluster (for $n_{cl} = 20$) is obtained. It is obvious that the indicated value N_{at} corresponds well to the adduced above nanoparticle definition criterion ($N_{at} = 10^3 \div 10^4$) [9, 17].

Let us consider synergetics of nanoclusters formation in PC and PAr. Using in the equation (3) as governing parameter critical magnitudes n_{cl} values at testing temperature T consecutive change and the indicated above the table of the determined by gold proportion law values A_m, m and Δ_i, the dependence $\Delta(T)$ can be obtained, which is adduced in Fig. 10.3. As it follows from this figure data, the nanoclusters stability within the temperature range of $313 \div 393$K is approximately constant and small ($\Delta_i \approx 0.232$ at minimum value $\Delta_i \approx 0.213$) and at $T > 393$K fast growth Δ_i (nanoclusters stability enhancement) begins for both considered polymers.

This plot can be explained within the frameworks of a cluster model [3–5]. In Fig. 10.3, glass transition temperatures of loosely packed matrix T_g', which are approximately 50 K lower than polymer macroscopic glass transition temperature T_g, are indicated by vertical shaded lines. At T_g' instable nanoclusters, that is, having small n_{cl} decay occurs. At the same time stable and, hence, more steady nanoclusters remain as a structural element, that results to Δ_i growth [14].

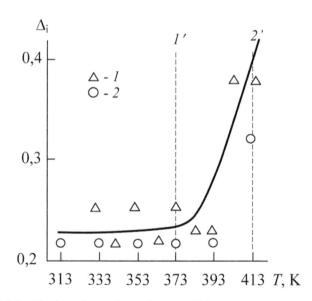

FIGURE 10.3   The dependence of nanoclusters stability measure Δ_i on testing temperature T for (1) PC and (2) PAr. The vertical shaded lines indicate temperature T_g' for PC (1') and PAR (2') [14].

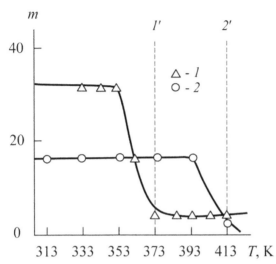

FIGURE 10.4   The dependences of reformations number m for nanoclusters on testing temperature T. The designations are the same as in Fig. 10.3 [14].

In Fig. 10.4, the dependences of reformations number m on testing temperature T for PC and PAr are adduced. At relatively low temperatures ($T < T_g'$) segments number in nanoclusters is large and segment joining (separation) to nanoclusters occurs easily enough, that explains large values m. At $T \rightarrow T_g'$ reformations number reduces sharply and at $T > T_g'$ $m \approx 4$. Since at $T > T_g'$ in the system only stable clusters remain, then it is necessary to assume, that large m at $T < T_g'$ are due to reformation of just instable nanoclusters [15].

In Fig. 10.5, the dependence of n_{cl} on m is adduced. As one can see, even small m enhancement within the range of 2÷16 results to sharp increasing in segments number per one nanocluster. At $m \approx 32$ the dependence $n_{cl}(m)$ attains asymptotic branch for both studied polymers. This supposes that $n_{cl} \approx 16$ is the greatest magnitude for nanoclusters and for $m \geq 32$ this term belongs equally to both joining and separation of such segment from nanocluster.

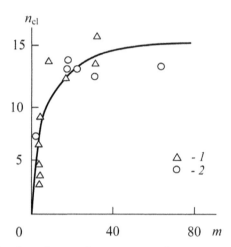

FIGURE 10.5 The dependence of segments number per one nanocluster n_{cl} on reformations number m for (1) PC and (2) PAr [14].

In Fig. 10.6, the relationship of stability measure Δ_i and reformations number m for nanoclusters in PC and PAr is adduced. As it follows from the data of this figure, at $m \geq 16$ (or, according to the data of Fig. 10.5, $n_{cl} \geq 12$) Δ_i value attains its minimum asymptotic magnitude $\Delta_i = 0.213$ [12]. This means, that for the indicated n_{cl} values nanoclusters in PC and PAr structure are adopted well to the external influence change ($A_m \geq 0.91$).

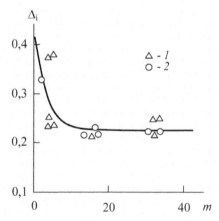

FIGURE 10.6   The dependence of stability measure Δ_i on reformation number m for PC (1) and PAR (2) [14]

Nanoclusters formation synergetics is directly connected with the studied polymers structure macroscopic characteristics. As it has been noted above, the fractal structure, characterized by the dimension d_f, is formed as a result of nanoclusters reformations. In Fig. 10.7, the dependence $d_f(\Delta_i)$ for the considered polymers is adduced, from which d_f increase at Δ_i growth follows. This means, that the increasing of possible reformations number m, resulting to Δ_i reduction (Fig. 10.6), defines the growth of segments number in nanoclusters, the latter relative fraction φ_{cl} enhancement and, as consequence, d_f reduction [3–5].

And let us note in conclusion the following aspect, obtaining from the plot $\Delta_i(T)$ (Fig. 10.3). extrapolation to maximum magnitude $\Delta_i \approx 1.0$. The indicated Δ_i value is reached approximately at $T \approx 458$ K that corresponds to mean glass transition temperature for PC and Par. Within the frameworks of the cluster model T_g reaching means polymer nanocluster structure decay [3–5] and, in its turn, realization at T_g of the condition $\Delta_i \approx 1.0$ means, that the "degenerated" nanocluster, consisting of one statistical segment or simply statistical segment, possesses the greatest stability measure. Several such segments joining up in nanocluster mains its stability reduction (see Figs. 10.5 and 10.6), that is the cause of glassy polymers structure thermodynamically nonequilibrium [14].

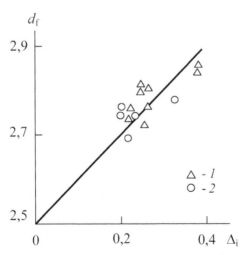

FIGURE 10.7   The dependence of structure fractal dimension d_f on stability measure of nanoclusters Δ_i for (1) PC and (2) PAr [14].

Therefore, the stated above results showed synergetics principles applicability for the description of association (dissociation) processes of polymer segments in local order domains (nanoclusters) in case of amorphous glassy polymers. Such conclusion can be a priori, since a nanoclusters are dissipative structures [6]. Testing temperature increase rises nanoclusters stability measure at the expense of possible reformations number reduction [14, 15].

As it has been shown lately, the notion "nanoparticle" (nanocluster) gets well over the limits of purely dimensional definition and means substance state specific character in sizes nanoscale. The nanoparticles, sizes of which are within the range of order of $1 \div 100$ nm, are already not classical macroscopic objects. They represent themselves the boundary state between macro and microworld and in virtue of this they have specific features number, to which the following ones are attributed:

1. nanoparticles are self-organizing nonequilibrium structures, which submit to synergetics laws;
2. they possess very mature surface;
3. nanoparticles possess quantum (wave) properties.

For the nanoworld structures in the form of nanoparticles (nanoclusters) their size, defining the surface energy critical level, is the information parameter of feedback [19].

The first from the indicated points was considered in detail above. The authors [20, 21] showed that nanoclusters surface fractal dimension changes within the range of $2.15 \div 2.85$ that is their well developed surface sign. And at last, let us consider quantum (wave) aspect of nanoclusters nature on the example of PC [22]. Structural levels hierarchy formation and development "scenario" in this case can be presented with the aid of iterated process [23]:

$$l_k = \langle a \rangle B_\lambda^k; \ \lambda_k = \langle a \rangle B_\lambda^{k+1}; \ k = 0, 1, 2, ..., \tag{4}$$

where l_k is specific spatial scale of structural changes, λ_k is length of irradiation sequence, which is due to structure reformation, k is structural hierarchy sublevel number, $B_\lambda = \lambda_b / \langle a \rangle = 2.61$ is discretely wave criterion of microfracture, λ_b is the smallest length of acoustic irradiation sequence.

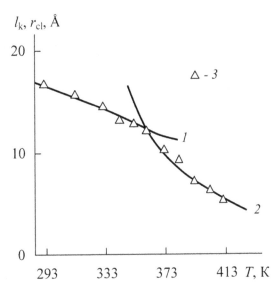

FIGURE 10.8   The dependences of structural changes specific spatial scale l_k at $B_\lambda = 1.06$ (1) and 1.19 (2) and nanoclusters radius r_{cl} (3) on testing temperature T for PC [22].

In Fig. 10.8, the dependences of l_k and nanoclusters radius r_{cl} on T are adduced, where l_k was determined according to the Eq. (4) and the value r_{cl} was calculated according to the formula (in Chapter 9). As it follows from the data of Fig. 10.8, the values l_k and r_{cl} agree within the whole studied temperatures range. Let us note, that if in paper [23] the value $B_\lambda = 2.61$, then for PC the indicated above agreement was obtained at $B_\lambda = 1.19$ and 1.06. This distinction confirms the thesis about distinction of synergetics laws in reference to nano-microworld objects (let us remind, that the condition $B_\lambda = 2.61$ is valid even in the case of earthquakes [14]). It is interesting to note, that B_λ change occurs at glass transition temperature of loosely packed matrix, that is, approximately at $T_g - 50$ K [11].

Hence, the stated above results demonstrated that the nanocluster possessed all nanoparticles properties, that is, they belonged to substance intermediate state-nanoworld.

And in completion of the present section let us note one more important feature of natural nanocomposites structure. In Refs. [24, 25], the interfacial regions absence in amorphous glassy polymers, treated as natural nanocomposites, was shown. This means, that such nanocomposites structure represents a nanofiller (nanoclusters), immersed in matrix (loosely packed matrix of amorphous polymer structure), that is, unlike polymer nano-composites with inorganic nanofiller (artificial nanocomposites) they have only two structural components.

10.3   THE NATURAL NANOCOMPOSITES REINFORCEMENT

As it is well-known [26], very often a filler introduction in polymer matrix is carried out for the last stiffness enhancement. Therefore the reinforce-ment degree of polymer composites, defined as a composite and matrix polymer elasticity moduli ratio, is one of their most important character-istics.

At amorphous glassy polymers as natural nanocomposites treatment the estimation of filling degree or nanoclusters relative fraction φ_{cl} has an important significance. Therefore, the authors [27] carried out the com-parison of the indicated parameter estimation different methods, one of which is EPR-spectroscopy (the method of spin probes). The indicated method allows to study amorphous polymer structural heterogeneity, us-ing radicals distribution character. As it is known [28], the method, based on the parameter d_1/d_c—the ratio of spectrum extreme components total intensity to central component intensity-measurement is the simplest and most suitable method of nitroxil radicals local concentrations determina-tion. The value of dipole–dipole interaction ΔH_{dd} is directly proportional to spin probes concentration C_w [29]:

$$\Delta H_{dd} = A \times C_w, \tag{5}$$

where $A = 5 \times 10^{-20}$ Ersted$\times$cm^3 in the case of radicals chaotic distribution.

On the basis of the Eq. (5) the relationship was obtained, which allows to calculate the average distance r between two paramagnetic probes [29]:

$$r = 38\left(\Delta H_{dd}\right)^{-1/3}, \text{Å} \tag{6}$$

where ΔH_{dd} is given in Ersteds.

In Fig. 10.9, the dependence of d_1/d_c on mean distance r between cha-otically distributed in amorphous PC radicals-probes is adduced. For PC at $T = 77K$ the values of $d_1/d_c = 0.38 \div 0.40$ were obtained. One can make an assumption about volume fractions relation for the ordered domains (nanoclusters) and loosely packed matrix of amorphous PC. The indicated value d_1/d_c means, that in PC at probes statistical distribution 0,40 of its volume is accessible for radicals and approximately 0.60 of volume remains unoccupied by spin probes, that is, the nanoclusters relative fraction φ_{cl} according to the EPR method makes up approximately $0.60 \div 0.62$.

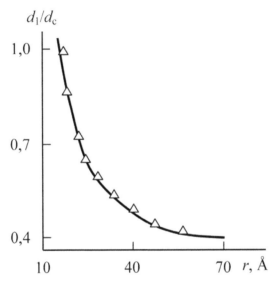

FIGURE 10.9 The dependence of parameter d_1/d_c of EPR spectrum on the value of mean distance r between radicals for PC [27].

This result corresponds well to the experimental data of Yech [30] and Perepechko [31], who obtained the values 0,60 and 0,63 for densely packed regions relative fraction in amorphous polymers.

The authors of Ref. [11] fulfilled φ_{cl} estimation with the aid of reversed gas chromatography and obtained the following magnitudes of this parameter for PC, poly (methyl methacrylate) and polysulfone: 0,70, 0,60 and 0,65, accordingly (Table 10.2).

Within the frameworks of the cluster model φ_{cl} estimation can be fulfilled by the percolation relationship (in Chapter 9) usage. Let us note, that in the given case the temperature of polymers structure quasiequilibrium state attainment, lower of which φ_{cl} value does not change, that is, T_0 [32], is accepted as testing temperature T. The calculation φ_{cl} results according to the equation (in Chapter 9) for the mentioned above polymers are adduced in Table 10.2, which correspond well to other authors estimations.

Proceeding from the circumstance, that radicals-probes are concentrated mainly in intercluster regions, the nanocluster size can be estimated, which in amorphous PC should be approximately equal to mean distance r between two paramagnetic probes, that is, ~ 50 Å (Fig. 10.9). This value corresponds well to the experimental data, obtained by dark-field electron microscopy method ($\approx 30 \div 100$ Å) [33].

Within the frameworks of the cluster model the distance between two neighboring nanoclusters can be estimated according to the equation (in previous paper) as $2R_{cl}$. The estimation $2R_{cl}$ by this mode gives the value 53,1 Å (at F = 41) that corresponds excellently to the method EPR data.

Thus, the Ref. [27] results showed, that the obtained by EPR method natural nanocomposites (amorphous glassy polymers) structure characteristics corresponded completely to both the cluster model theoretical calculations and other authors estimations. In other words, EPR data are experimental confirmation of the cluster model of polymers amorphous state structure.

The treatment of amorphous glassy polymers as natural nanocomposites allows to use for their elasticity modulus E_p (and, hence, the reinforcement degree $E_p/E_{l.m.}$, where $E_{l.m.}$ is loosely packed matrix elasticity modulus) description theories, developed for polymer composites reinforcement degree description [9, 17]. The authors [34] showed correctness of particulate-filled polymer nanocomposites reinforcement of two concepts on the example of amorphous PC. For theoretical estimation of particulate-filled polymer nanocomposites reinforcement degree E_n/E_m two equations can be used. The first from them has the look [35]:

$$\frac{E_n}{E_m} = 1 + \phi_n^{1,7}, \tag{7}$$

where E_n and E_m are elasticity moduli of nanocomposites and matrix polymer, accordingly, φ_n is nanofiller volume contents.

The second equation offered by the authors of Ref. [36] is:

$$\frac{E_n}{E_m} = 1 + \frac{0,19 W_n l_{st}}{D_p^{1/2}},$$

(8)

where W_n is nanofiller mass contents in mass.%, D_p is nanofiller particles diameter in nm.

Let us consider included in the Eqs. (7) and (8) parameters estimation methods. It is obvious, that in the case of natural nanocomposites one should accept: $E_n = E_p$, $E_m = E_{l.m.}$ and $\varphi_n = \varphi_{cl}$, the value of the latter can be estimated according to the equation (in Chapter 9).

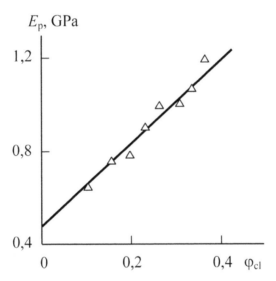

FIGURE 10.10 The dependence of elasticity modulus E_p on nanoclusters relative fraction φ_{cl} for PC [34].

The mass fraction of nanoclusters W_{cl} can be calculated as follows [37]:

$$W_{cl} = \rho \varphi_{cl},$$

(9)

where ρ is nanofiller (nanoclusters) density, which is equal to 1300 kg/m³ for PC.

The value $E_{l.m.}$ can be determined by the construction of $E_p(\varphi_{cl})$ plotting, which is adduced in Fig. 10.10. As one can see, this plot is approximately linear and its extrapolation to $\varphi_{cl} = 0$ gives the value $E_{l.m.}$ And at last, as it follows from the nanoclusters definition (*see* Chapter 1) one should accept $D_p \approx l_{st}$ for them and then the equation (8) accepts the following look [34]:

In Fig 10.11, the comparison of theoretical calculation according to the Eqs. (7) and (10) with experimental values of reinforcement degree $E_p/E_{1.m.}$ for PC is adduced. As one can see, both indicated equations give a good enough correspondence with the experiment: their average discrepancy makes up 5.6% in the Eq. (7) case and 9.6% for the Eq. (10). In other words, in both cases the average discrepancy does not exceed an experimental error for mechanical tests. This means, that both considered methods could be used for PC elasticity modulus prediction. Besides, it necessary to note, that the percolation Eq. (7) qualitatively describes the dependence $E_p/E_{1.m.}$ (φ_{cl}) better, than the empirical Eq. (10).

$$\frac{E_n}{E_m} = 1 + 0,19\rho\phi_{cl}l_{st}^{1/2} \ . \tag{10}$$

The obtained results allowed to make another important conclusion. As it is known, the percolation Eq. (7) assumes, that nanofiller is percolation system (polymer composite) solid-body component and in virtue of this circumstance defines this system elasticity modulus. However, for artificial polymer particulate-filled nanocomposites, consisting of polymer matrix and inorganic nanofiller, the Eq. (7) in the cited form gives the understated values of reinforcement degree. The authors [9, 17] showed, that for such nanocomposites the sum $(\varphi_n+\varphi_{if})$, where φ_{if} was interfacial regions relative fraction, was a solid-body component. The correspondence of experimental data and calculation according to the Eq. (7) demonstrates, that amorphous polymer is the specific nanocomposite, in which interfacial regions are absent [24, 25]. This important circumstance is necessary to take into consideration at amorphous glassy polymers structure and properties description while simulating them as natural nanocomposites. Besides, one should note, that unlike micromechanical models the Eqs. (7) and (10) do not take into account nanofiller elasticity modulus, which is substantially differed for PC nanoclusters and inorganic nanofillers [34].

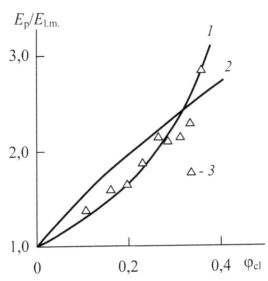

FIGURE 10.11 The dependences of reinforcement degree $E_p/E_{1.m}$ on nanoclusters relative fraction φ_{cl} for PC. 1 – calculation according to the Eq. (7); 2 – calculation according to the Eq. (10); 3 – the experimental data [34].

Another mode of natural nanocomposites reinforcement degree description is micromechanical models application, developed for polymer composites mechanical behavior description [1, 37–39]. So, Takayanagi and Kerner models are often used for the description of reinforcement degree on composition for the indicated materials [38, 39]. The authors of Ref. [40] used the mentioned models for theoretical treatment of natural nanocomposites reinforcement degree temperature dependence on the example of PC.

Takayanagi model belongs to a micromechanical composite models group, allowing empirical description of composite response upon mechanical influence on the basis of constituent it elements properties. One of the possible expressions within the frameworks of this model has the following look [38]:

$$\frac{G_c}{G_m} = \frac{\phi_m G_m + (\alpha + \phi_f) G_f}{(1 + \alpha \phi_f) G_m + \alpha \phi_m G_f}, \qquad (11)$$

where G_c, G_m and G_f are shear moduli of composite, polymer matrix and filler, accordingly, φ_m and φ_f are polymer matrix and filler relative fractions, respectively, α is a fitted parameter.

Kerner equation is identical to the Eq. (11), but for it the parameter α does not fit and has the following analytical expression [38]:

$$\alpha_m = \frac{2(4-5v_m)}{(7-5v_m)} \tag{12}$$

where α_m and v_m are parameter α and Poisson's ratio for polymer matrix.

Let us consider determination methods of the Eqs. (11) and (12) parameters, which are necessary for the indicated equations application in the case of natural nanocomposites, Firstly, it is obvious, that in the last case one should accept: $G_c = G_p$, $G_m = G_{l.m.}$, $G_f = G_{cl}$, where G_p, $G_{l.m.}$ and G_{cl} are shear moduli of polymer, loosely packed matrix and nanoclusters, accordingly, and also $\varphi_f = \varphi_{cl}$, where φ_{cl} is determined according to the percolation relationship (in Chapter 9). Young's modulus for loosely packed matrix and nanoclusters can be received from the data of Fig. 10.10 by the dependence $E_p(\varphi_{cl})$ extrapolation to $\varphi_{cl} = 1.0$, respectively. The corresponding shear moduli were calculated according to the general equation (in Chapter 9). The value of nanoclusters fractal dimension d_f^{cl} in virtue of their dense package is accepted equal to the greatest dimension for real solids ($d_f^{cl} = 2.95$ [40]) and loosely packed matrix fractal dimension $d_f^{l.m.}$ can be estimated.

However, the calculation according to the Eqs. (11) and (12) does not give a good correspondence to the experiment, especially for the temperature range of $T = 373 \div 413$ K in PC case. As it is known [38], in empirical modifications of Kerner equation it is usually supposed, that nominal concentration scale differs from mechanically effective filler fraction φ_f^ε, which can be written accounting for the designations used above for natural nanocomposites as follows [41].

$$\phi_f^{ef} = \frac{(G_p - G_{l.m.})(G_{l.m.} + \alpha_{l.m.}G_{cl})}{(G_{cl} - G_{l.m.})(G_{l.m.} + \alpha_{l.m.}G_p)}, \tag{13}$$

where $\alpha_{l.m.} = \alpha_m$. The value $\alpha_{l.m.}$ can be determined according to the Eq. (12), estimating Poisson's ratio of loosely packed matrix $v_{l.m.}$ by the known values $d_f^{l.m.}$ according to the equation (in Chapter 9).

Besides, one more empirical modification ϕ_f^{ef} exists, which can be written as follows [41]:

$$\phi_{cl_2}^{ef} = \phi_{cl} + c\left(\frac{\phi_{cl}}{2r_{cl}}\right)^{2/3},$$

(14)

where c is empirical coefficient of order one r_{cl} is nanocluster radius, determined according to the equation (in Chapter 9).

At the value $\phi_{cl_2}^{ef}$ calculation according to the Eq. (14) magnitude c was accepted equal to 1.0 for the temperature range of $T = 293 \div 363$ K and equal to 1.2, for the range of $T = 373 \div 413$ K and $2r_{cl}$ is given in nm. In Fig. 10.12, the comparison of values ϕ_{cl}^{ef}, calculated according to the equations (13) and (14) ($\phi_{cl_1}^{ef}$ and $\phi_{cl_2}^{ef}$, accordingly) is adduced. As one can see, a good enough conformity of the values ϕ_{cl}^{ef}, estimated by both methods, is obtained (the average discrepancy of $\phi_{cl_1}^{ef}$ and $\phi_{cl_2}^{ef}$ makes up slightly larger than 20%). Let us note, that the effective value φ_{cl} exceeds essentially the nominal one, determined according to the relationship (in Chapter 9): within the range of $T = 293 \div 363$K by about 70% and within the range of $T = 373 \div 413$ K, almost in three times.

In Fig. 10.13, the comparison of experimental and calculated according to Kerner equation (the Eq. (11)) with the Eqs. (13) and (14) using values of reinforcement degree by shear modulus $G_p/G_{l.m.}$ as a function of testing temperature T for PC is adduced. As one can see, in this case at the usage of nanoclusters effective concentration scale (ϕ_{cl}^{ef} instead of φ_{cl}) the good conformity of theory and experiment is obtained (their average discrepancy makes up 6%).

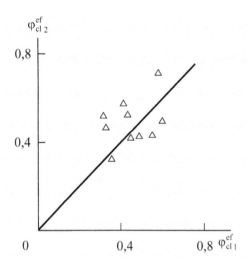

FIGURE 10.12   The comparison of nanoclusters effective concentration scale $\phi_{cl_1}^{ef}$ and $\phi_{cl_2}^{ef}$, calculated according to the Eqs. (13) and (14), respectively, for PC. A straight line shows the relation 1:1 [41].

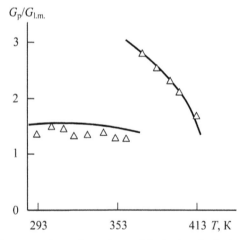

FIGURE 10.13   The comparison of experimental (points) and calculated according to the Eqs. (11), (13) and (14) (solid lines) values of reinforcement degree by shear modulus $G_p/G_{l.m.}$ as a function of testing temperature T for PC [41].

Hence, the stated above results have shown the modified Kerner equation application correctness for natural nanocomposites elastic response description. Really this fact by itself confirms the possibility of amorphous glassy polymers treatment as nanocomposites. Microcomposite models usage gives the clear notion about factors, influencing polymers stiffness.

10.4 INTERCOMPONENT ADHESION IN NATURAL NANOCOMPOSITES

Amorphous glassy polymers as natural nanocomposites puts forward to the foreground their study intercomponent interactions, that is, interactions nanoclusters – loosely packed matrix. This problem plays always one of the main roles at multiphase (multicomponent) systems consideration, since the indicated interactions or interfacial adhesion level defines to a great extent such systems properties [42]. Therefore, the authors [43] studied the physical principles of intercomponent adhesion for natural nanocomposites on the example of PC.

The authors [44] considered three main cases of the dependence of reinforcement degree E_c/E_m on φ_f. In this chapter, the authors have shown, that there are the following main types of the dependences $E_c/E_m(\varphi_f)$ exist:

1. the ideal adhesion between filler and polymer matrix, described by Kerner equation (perfect adhesion), which can be approximated by the following relationship:

$$\frac{E_c}{E_m} = 1 + 11,64\phi_f - 44,4\phi_f^2 + 96,3\phi_f^3 ; \qquad (15)$$

2. zero adhesional strength at a large friction coefficient between filler and polymer matrix, which is described by the equation:

$$\frac{E_c}{E_m} = 1 + \phi_f ; \qquad (16)$$

3. the complete absence of interaction and ideal slippage between filler and polymer matrix, when composite elasticity modulus is defined practically by polymer cross-section and connected with the filling degree by the equation:

$$\frac{E_c}{E_m} = 1 - \phi_f^{2/3} \cdot \qquad (17)$$

In Fig. 10.14 ,the theoretical dependences $E_p/E_{1.m.}(\varphi_{cl})$ plotted according to the Eqs. (15) ÷ (17), as well as experimental data (points) for PC are shown. As it follows from the adduced in Fig. 10.14 comparison at $T = 293÷363$ K the experimental data correspond well to the Eq. (16), that is, in this case zero adhesional strength at a large friction coefficient is observed. At $T = 373÷413$ K the experimental data correspond to the Eq. (15), that is, the perfect adhesion between nanoclusters and loosely packed matrix is observed. Thus, the adduced in Fig. 10.14 data demonstrated, that depending on testing temperature two types of interactions nanoclusters – loosely packed matrix are observed: either perfect adhesion or large friction between them. For quantitative estimation of these interactions it is necessary to determine their level, which can be made with the help of the parameter b_m, which is determined according to the equation [45]:

$$\sigma_f^c = \sigma_f^m K_s - b_m \phi_f , \tag{17}$$

where σ_f^c and σ_f^m are fracture stress of composite and polymer matrix, respectively, K_s is stress concentration coefficient. It is obvious, that since b_m increase results to σ_f^c reduction, then this means interfacial adhesion level decrease.

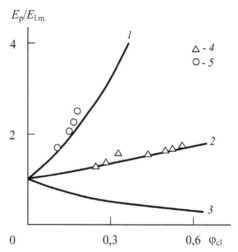

FIGURE 10.14 The dependences of reinforcement degree $E_p/E_{1.m}$ on nanoclusters relative fraction φ_{cl}. 1÷3—the theoretical dependences, corresponding to the Eqs. (15) ÷ (17), accordingly; 4, 5—the experimental data for PC within the temperature ranges: 293÷363 K (4) and 373÷413 K (5) [43].

The true fracture stress σ_f^{tr} for PC, taking into account sample cross-section change in a deformation process, was used as σ_f^c for natural nano-composites, which can be determined according to the known formula:

$$\sigma_f^{tr} = \sigma_f^n \left(1 + \varepsilon_f\right),$$ (19)

where σ_f^n is nominal (engineering) fracture stress, ε_f is strain at fracture.

The value σ_f^m, which is accepted equal to loosely packed matrix strength $\sigma_f^{l.m.}$, was determined by graphic method, namely, by the dependence $\sigma_f^t (\varphi_{cl})$ plotting, which proves to be linear, and by subsequent extrapolation of it to $\varphi_{cl} = 0$, that gives $\sigma_f^{l.m.} = 40$ MPa [43].

And at last, the value K_s can be determined with the help of the following equation [39]:

$$\sigma_f^{tr} = \sigma_f^{l.m.} \left(1 - \phi_{cl}^{2/3}\right) K_s .$$ (20)

The parameter b_m calculation according to the stated above technique shows its decrease (intercomponent adhesion level enhancement) at testing temperature raising within the range of $b_m \approx 500 \div 130$.

For interactions nanoclusters – loosely packed matrix estimation within the range of $T = 293 \div 373$ K the authors [48] used the model of Witten-Sander clusters friction, stated in Ref. [46]. This model application is due to the circumstance, that amorphous glassy polymer structure can be presented as an indicated clusters large number set [47]. According to this model, Witten-Sander clusters generalized friction coefficient t can be written as follows [46]:

$$f = \ln c + \beta \times \ln n_{cl},$$ (21)

where c is constant, β is coefficient, n_{cl} is statistical segments number per one nanocluster.

The coefficient β value is determined as follows [46]:

$$\beta = \left(d_f^{cl}\right)^{-1},$$ (22)

where d_f^{cl} is nanocluster structure fractal dimension, which is equal, as before, to 2.95 [40].

In Fig. 10.15, the dependence $b_m(f)$ is adduced, which is broken down into two parts. On the first of them, corresponding to the range of $T = 293 \div 363$ K, the intercomponent interaction level is intensified at f decreasing (i.e., b_m reduction is observed and on the second one, corresponding to the range of $T = 373 \div 413$ K, $b_m = $ const independent on value f. These results correspond completely to the data of Fig. 10.14, where in the first from the indicated temperature ranges the value $E_p/E_{1.m.}$ is defined by nano-clusters friction and in the second one by adhesion and, hence, it does not depend on friction coefficient.

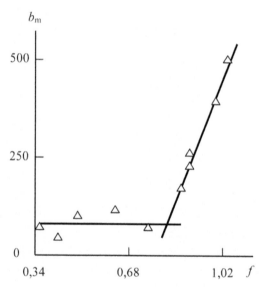

FIGURE 10.15   The dependence of parameter b_m on generalized friction coefficient f for PC [43].

As it has been shown in Ref. [48], the interfacial (or intercomponent) adhesion level depends on a number of accessible for the formation interfacial (intercomponent) bond sites (nodes) on the filler (nanocluster) particle surface N_u, which is determined as follows [49]:

$$N_u = L^{d_u},$$
(23)

where L is filler particle size, d_u is fractal dimension of accessible for contact ("nonscreened") indicated particle surface.

One should choose the nanocluster characteristic size as L for the natural nanocomposite which is equal to statistical segment l_{st}, determined according to the equation (in Chapter 9), and the dimension d_u is determined according to the following relationship [49]:

$$d_u = (d_{surf} - 1) + \left(\frac{d - d_{surf}}{d_w}\right),\tag{24}$$

where d_{surf} is nanocluster surface fractal dimension, d_w is dimension of random walk on this surface, estimated according to Aarony-Stauffer rule [49]:

$$d_w = d_{surf} + 1.\tag{25}$$

The following technique was used for the dimension d_{surf} calculation. First the nanocluster diameter $D_{cl} = 2r_{cl}$ was determined according to the equation (in Chapter 9) and then its specific surface S_u was estimated [35]:

$$S_u = \frac{6}{\rho_{cl} D_{cl}},\tag{26}$$

where ρ_{cl} is the nanocluster density, equal to 1300 kg/m³ in the PC case.

And at last, the dimension d_{surf} was calculated with the help of the equation [20]:

$$S_u = 5,25 \times 10^3 \left(\frac{D_{cl}}{2}\right)^{d_{surf} - d}.\tag{27}$$

In Fig. 10.16, the dependence $b_m(N_u)$ for PC is adduced, which is broken down into two parts similarly to the dependence $b_m(f)$ (Fig. 10.15). At $T = 293 \div 363$ K the value b_m is independent on N_u, since nanocluster – loosely packed matrix interactions are defined by their friction coefficient. Within the range of $T = 373 \div 413$ K intercomponent adhesion level enhancement (b_m reduction) at active sites number N_u growth is observed, as was to be expected. Thus, the data of both Figs. 10.15 and 10.16 correspond to Fig. 10.14 results.

With regard to the data of Figs. 10.15 and 10.16, two remarks should be made. Firstly, the transition from one reinforcement mechanism to another corresponds to loosely packed matrix glass transition temperature, which is approximately equal to $T_g - 50$ K [11]. Secondly, the extrapolation of Fig. 10.16 plot to $b_m = 0$ gives the value $N_u \approx 71$, that corresponds approximately to polymer structure dimension $d_f = 2.86$.

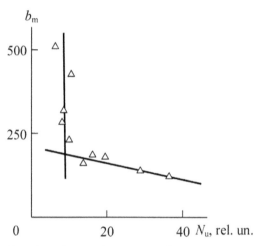

FIGURE 10.16 The dependence of parameter b_m on nanocluster surface active ("nonscreened") sites number N_u for PC [43].

In this theme completion an interesting structural aspect of intercomponent adhesion in natural nanocomposites (polymers) should be noted. Despite the considered above different mechanisms of reinforcement and nanoclusters-loosely packed matrix interaction realization the common dependence $b_m(\varphi_{cl})$ is obtained for the entire studied temperature range of 293÷413 K, which is shown in Fig. 10.17. This dependence is linear, that allows to determine the limiting values $b_m \approx 970$ at $\varphi_{cl} = 1.0$ and $b_m = 0$ at $\varphi_{cl} = 0$. Besides, let us note, that the shown in Figs. 10.14÷10.16 structural transition is realized at $\varphi_{cl} \approx 0.26$ [43].

Hence, the stated above results have demonstrated, that intercomponent adhesion level in natural nanocomposites (polymers) has structural origin and is defined by nanoclusters relative fraction. In two temperature ranges two different reinforcement mechanisms are realized, which are due to large friction between nanoclusters and loosely packed matrix and also perfect (by Kerner) adhesion between them. These mechanisms can be described successfully within the frameworks of fractal analysis.

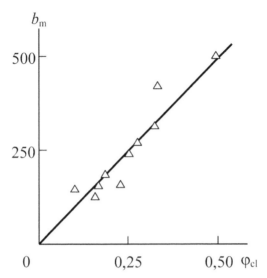

FIGURE 10.17   The dependence of parameter b_m on nanoclusters relative fraction φ_{cl} for PC [43].

The further study of intercomponent adhesion in natural nanocomposites was fulfilled in Ref. [50]. In Fig. 10.18, the dependence $b_m(T)$ for PC is shown, from which b_m reduction or intercomponent adhesion level enhancement at testing temperature growth follows. In the same figure the maximum value b_m for nanocomposites polypropylene/Na^+-montmorillonite [9] was shown by a horizontal shaded line. As one can see, b_m values for PC within the temperature range of $T = 373 \div 413$ K by absolute value are close to the corresponding parameter for the indicated nanocomposite, that indicates high enough intercomponent adhesion level for PC within this temperature range.

Let us note an important structural aspect of the dependence $b_m(T)$, shown in Fig. 10.18. According to the cluster model [4], the decay of instable nanoclusters occurs at temperature $T'_g \approx T_g - 50$ K, holding back loosely packed matrix in glassy state, owing to which this structural component is devitrificated within the temperature range of $T'_g \div T_g$. Such effect results to rapid reduction of polymer mechanical properties within the indicated temperature range [51]. As it follows from the data of Fig. 10.18, precisely in this temperature range the highest intercomponent adhesion level is observed and its value approaches to the corresponding character-istic for nanocomposites polypropylene/Na$^+$-montmorillonite.

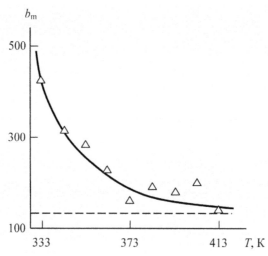

FIGURE 10.18   The dependence of parameter b_m on testing temperature T for PC. The horizontal shaded line shows the maximum value b_m for nanocomposites polypropylene/Na$^+$-montmorillonite [50].

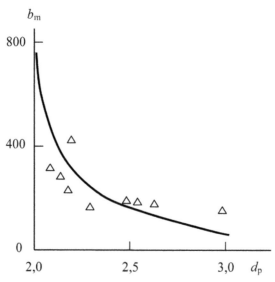

FIGURE 10.19   The dependence of parameter b_m on nanoclusters surface fractal dimension d_{surf} for PC [50].

It can be supposed with a high probability degree that adhesion level depends on the structure of nanoclusters surface, coming into contact with loosely packed matrix, which is characterized by the dimension d_{surf}. In Fig. 10.19, the dependence $b_m(d_{surf})$ for PC is adduced, from which rapid reduction b_m (or intercomponent adhesion level enhancement) follows at d_{surf} growth or, roughly speaking, at nanoclusters surface roughness enhancement.

The authors [48] showed that the interfacial adhesion level for composites polyhydroxyether/graphite was raised at the decrease of polymer matrix and filler particles surface fractal dimensions difference. The similar approach was used by the authors of Ref. [50], who calculated nanoclusters d_f^{cl} and loosely packed matrix $d_f^{l.m.}$ fractal dimensions difference Δd_f:

$$\Delta df = d_f^{cl} - d_f^{l.m.}, \tag{28}$$

where d_f^{cl} is accepted equal to real solids maximum dimension ($d_f^{cl} = 2.95$ [40]) in virtue of their dense packing and the value $d_f^{l.m.}$ was calculated according to the mixtures rule (the equation from Chapter 9).

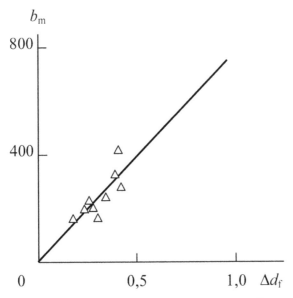

FIGURE 10.20   The dependence of parameter b_m on nanoclusters d_f^{cl} and loosely packed matrix $d_f^{l.m.}$ structures fractal dimensions difference Δd_f for PC [50]

In Fig. 10.20, the dependence of b_m on the difference Δd_f is adduced, from which b_m decrease or intercomponent adhesion level enhancement at Δd_f reduction or values d_f^{cl} and $d_f^{l.m.}$ growing similarity follows. This dependence demonstrates, that the greatest intercomponent adhesion level, corresponding to $b_m = 0$, is reached at $\Delta d_f = 0.95$ and is equal to ~ 780.

The data of Figs. 10.14 and 10.18 combination shows, that the value $b_m \approx 200$ corresponds to perfect adhesion by Kerner. In its turn, the Figs. 10.16 and 10.17 plots data demonstrated, that the value $b_m \approx 200$ could be obtained either at $d_{surf} > 2.5$ or at $\Delta d_f < 0,3$, accordingly. The obtained earlier results showed [24], that the condition $d_{surf} > 2.5$ was reached at $r_{cl} < 7.5$Å or $T > 373$ K, that again corresponded well to the stated above results. And at last, the $\Delta d_f \approx 0,3$ or $d_f^{l.m.} \approx 2.65$ according to the equation (in Chapter 9) was also obtained at $T \approx 373$ K.

Hence, at the indicated above conditions fulfillment within the temperature range of $T < T_g'$ for PC perfect intercomponent adhesion can be obtained, corresponding to Kerner equation, and then the value E_p estimation should be carried out according to the equation (15). At $T = 293$ K ($\varphi_{cl} = 0.56$, $E_m = 0.85$ GPa) the value E_p will be equal to 8.9 GPa, that approximately in 6 times larger, than the value E_p for serial industrial PC brands at the indicated temperature.

Let us note the practically important feature of the obtained above results. As it was shown, the perfect intercomponent adhesion corresponds to $b_m \approx 200$, but not $b_m = 0$. This means, that the real adhesion in natural nanocomposites can be higher than the perfect one by Kerner, which was shown experimentally on the example of particulate-filled polymer nanocomposites [17, 52]. This effect was named as nanoadhesion and its realization gives large possibilities for elasticity modulus increase of both natural and artificial nanocomposites. So, the introduction in aromatic polyamide (phenylone) of 0.3 mass.% aerosil only at nanoadhesion availability gives the same nanocomposite elasticity modulus enhancement effect, as the introduction of 3 mass.% of organoclay, which at present is assumed as one of the most effective nanofillers [9]. This assumes, that the value $E_p = 8.9$ GPa for PC is not a limiting one, at any rate, theoretically. Let us note in addition, that the indicated E_p values can be obtained at the natural nanocomposites nanofiller (nanoclusters) elasticity modulus magnitude $E_{cl} = 2.0$ GPa, that is, at the condition $E_{cl} < E_p$. Such result possibility follows from the polymer composites structure fractal concept [53], namely, the model [44], in which the equations (15) ÷ (17) do not contain nanofiller elasticity modulus, and reinforcement percolation model [35].

The condition d_{surf} < 2.5, that is, r_{cl} < 7.5 Å or N_{cl} < 5, in practice can be realized by the nanosystems mechanosynthesis principles using, the grounds of which are stated in Ref. [54]. However, another more simple and, hence, more technological method of desirable structure attainment realization is possible, one from which will be considered in subsequent section.

Hence, the stated above results demonstrated, that the adhesion level between natural nanocomposite structural components depended on nanoclusters and loosely packed matrix structures closeness. This level change can result to polymer elasticity modulus significant increase. A number of these effect practical realization methods were considered [50].

The mentioned above dependence of intercomponent adhesion level on nanoclusters radius r_{cl} assumes more general dependence of parameter b_m on nanoclusters geometry. The authors [55] carried out calculation of accessible for contact sites of nanoclusters surface and loosely packed matrix number N_u according to the Eq. (23) for two cases. The nanocluster is simulated as a cylinder with diameter D_{cl} and length l_{st}, where l_{st} is statistical segment length, therefore, in the first case its butt-end is contacting with loosely packed matrix nanocluster surface and then $L = D_{cl}$ and in the second case with its side (cylindrical) surface and then $L = l_{st}$. In Fig. 10.21, the dependences of parameter b_m on value N_u, corresponding to the two considered above cases, are adduced. As one can see, in both cases, for the range of $T = 293 ÷ 363$ K l_{st}, where interactions nanoclusters − loosely packed matrix are characterized by powerful friction between them, the value b_m does not depend on N_u, as it was expected. For the range of $T = 373 ÷ 413$ K, where between nanoclusters and loosely packed matrix perfect adhesion is observed, the linear dependences $b_m(N_u)$ are obtained. However, at using value D_{cl} as Lb_m reduction or intercomponent adhesion level enhancement at N_u decreasing is obtained and at $N_u = 0$, b_m value reaches its minimum magnitude $b_m = 0$. In other words, in this case the minimum level of intercomponent adhesion is reached at intercomponent bonds formation sites (nodes) absence that is physically incorrect [48]. And on the contrary at the condition $L = l_{st}b_m$ the reduction (intercomponent adhesion level enhancement) at the increase of contacts number N_u between nanoclusters and loosely packed matrix is observed, that is obvious from the physical point of view. Thus, the data of Fig. 10.21 indicate unequivocally, that the intercomponent adhesion is realized over side (cylindrical) nanoclusters surface and butt-end surfaces in this effect formation do not participate.

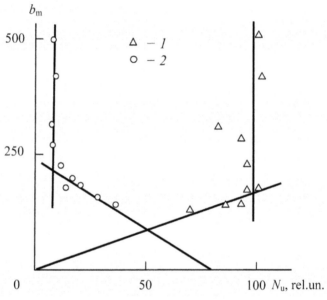

FIGURE 10.21 The dependences of parameter b_m on a number of accessible for intercomponent bonds formation sizes on nanocluster surface N_u at the condition $L = D_{cl}$ (1) and $L = l_{st}$ (2) for PC [55].

Let us consider geometrical aspects intercomponent interactions in natural nanocomposites. In Fig. 10.22, the dependence of nanoclusters butt-end S_b and side (cylindrical) S_c surfaces areas on testing temperature T for PC are adduced. As one can see, the following criterion corresponds to the transition from strong friction to perfect adhesion at $T = 373K$ [55]:

$$S_b \approx S_c. \tag{29}$$

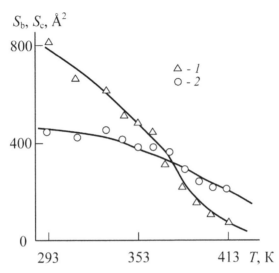

FIGURE 10.22 The dependences of nanoclusters butt-end S_b (1) and cylindrical S_c (2) surfaces areas on testing temperature T for PC [55].

Hence, the intercomponent interaction type transition from the large friction nanoclusters – loosely packed matrix to the perfect adhesion between them is defined by nanoclusters geometry: at $S_b > S_c$ the interactions of the first type is realized and at $S_b < S_c$, the second one. Proceeding from this, it is expected that intercomponent interactions level is defined by the ratio S_b/S_c. Actually, the adduced in Fig. 10.23 data demonstrate b_m reduction at the indicated ratio decrease, but at the criterion (the Eq. (29)) realization or $S_b/S_c \approx 1$ S_b/S_c S_b/S_c decreasing does not result to b_m reduction and at $S_b/S_c < 1$ intercomponent adhesion level remains maximum high and constant [55].

Hence, the stated above results have demonstrated, that interactions nanoclusters-loosely packed matrix type (large friction or perfect adhesion) is defined by nanoclusters butt-end and side (cylindrical) surfaces areas ratio or their geometry if the first from the mentioned areas is larger that the second one then a large friction nanoclusters-loosely packed matrix is realized; if the second one exceeds the first one, then between the indicated structural components perfect adhesion is realized. In the second from the indicated cases intercomponent adhesion level does not depend on the mentioned areas ratio and remains maximum high and constant. In other words, the adhesion nanoclusters-loosely packed matrix is realized by nanoclusters cylindrical surface.

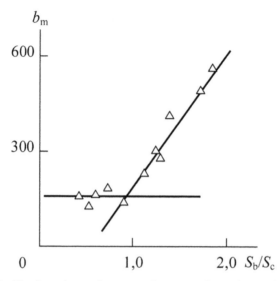

FIGURE 10.23   The dependence of parameter b_m on nanoclusters butt-end and cylindrical surfaces are ratio S_b/S_c value for PC [55].

The stated above results were experimentally confirmed by the EPR-spectroscopy method [56]. The Eqs. (1) and (6) comparison shows, that dipole–dipole interaction energy ΔH_{dd} has structural origin, namely [56]:

$$\Delta H_{dd} \approx \left(\frac{V_{cl}}{n_{cl}} \right). \tag{30}$$

As estimations according to the Eq. (30) showed, within the temperature range of $T = 293 \div 413K$ for PC ΔH_{dd} increasing from 0.118 up to 0.328 Ersteds was observed.

Let us consider dipole–dipole interaction energy ΔH_{dd} intercommunication with nanoclusters geometry. In Fig. 10.24, the dependence of ΔH_{dd} on the ratio S_c/S_b for PC is adduced. As one can see, the linear growth ΔH_{dd} at ratio S_c/S_b increasing is observed, that is, either at S_c enhancement or at S_b reduction. Such character of the adduced in Fig. 10.24 dependence indicates unequivocally, that the contact nanoclusters-loosely packed matrix is realized on nanocluster cylindrical surface. Such effect was to be expected, since emerging from the butt-end surface statistically distributed polymer chains complicated the indicated contact realization unlike relatively smooth cylindrical surfaces. It is natural to suppose, that dipole–dipole interactions intensification or ΔH_{dd} increasing results to natural nanocomposites elasticity modulus E_p enhancement. The second as natural supposition at PC consideration as nanocomposite is the influence on the value E_p of nanoclusters (nanofiller) relative fraction φ_{cl}, which is determined according to the percolation relationship (in Chapter 9).

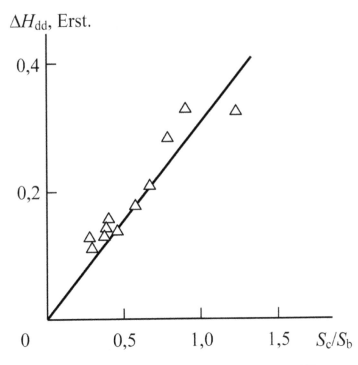

FIGURE 10.24 The dependence of dipole–dipole interaction energy ΔH_{dd} on nanoclusters cylindrical S_c and butt-end S_b surfaces areas ratio for PC [56].

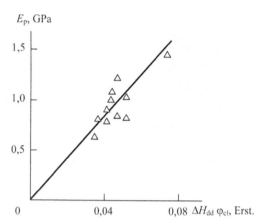

FIGURE 10.25   The dependence of elasticity modulus E_p on complex argument $(\Delta H_{dd}\varphi_{cl})$ for PC [56].

In Fig. 10.25, the dependence of elasticity modulus E_p on complex argument $(\Delta H_{dd}\varphi_{cl})$ for PC is presented. As one can see, this dependence is a linear one, passes through coordinates origin and is described analytically by the following empirical equation [56].

$$E_p = 21(\Delta H_{dd}\varphi_{cl}), \text{ GPa,} \qquad (31)$$

which with the appreciation of the equation (30) can be rewritten as follows [56]:

$$E_p = 21 \times 10^{-26} \left(\frac{\phi_{cl} i_{cl}}{n_{cl}} \right), \text{ GPa.} \qquad (32)$$

The Eq. (32) demonstrates clearly, that the value E_p and, hence polymer reinforcement degree is a function of its structural characteristics, described within the frameworks of the cluster model [3–5]. Let us note, that since parameters v_{cl} and φ_{cl} are a function of testing temperature, then the parameter n_{cl} is the most suitable factor for the value E_p regulation for practical purposes. In Fig. 10.26, the dependence $E_p(n_{cl})$ for PC at $T = 293$ K is adduced, calculated according to the Eq. (32), where the values v_{cl} and φ_{cl} were calculated according to the equations (in Chapter 9). As one can see, at small n_{cl} (<10) the sharp growth E_p is observed and at the smallest possible value $n_{cl} = 2$ the magnitude $E_p \approx 13.5$ GPa. Since for PC $E_{1m.} = 0.85$ GPa, then it gives the greatest reinforcement degree E_p/E_m 15.9. Let us note, that the greatest attainable reinforcement degree for artificial nanocomposites (polymers filled with inorganic nanofiller) cannot exceed 12 [9]. It is notable, that the shown in Fig. 10.26 dependence $E_p(n_{cl})$ for PC is identical completely by dependence shape to the dependence of elasticity modulus of nanofiller particles diameter for elastomeric nanocomposites [57].

Hence, the presented above results have shown that elasticity modulus of amorphous glassy polycarbonate, considered as natural nanocomposite, are defined completely by its suprasegmental structure state. This state can be described quantitatively within the frameworks of the cluster model of polymers amorphous state structure and characterized by local order level. Natural nanocomposites reinforcement degree can essentially exceed analogous parameter for artificial nanocomposites [56].

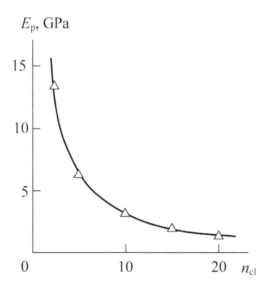

FIGURE 10.26   The dependence of elasticity modulus E_p on segments number n_{cl} per one nanocluster, calculated according to the Eq. (32) for PC at $T = 293$ K [56].

As it has been shown above (see the Eqs. (7) and (15)), the nanocluster relative fraction increasing results to polymers elasticity modulus enhancement similarly to nanofiller contents enhancement in artificial nanocomposites. Therefore the necessity of quantitative description and subsequent comparison of reinforcement degree for the two indicated above nanocomposites classes appears. The authors [58, 59] fulfilled the comparative analysis of reinforcement degree by nanoclusters and by layered silicate (organoclay) for polyarylate and nanocomposite epoxy polymer/Na$^+$ - montmorillonite [60], accordingly.

In Fig. 10.27, theoretical dependences of reinforcement degree E_n/E_m on nanofiller contents φ_n, calculated according to the Eqs. (15) ÷ (17), are adduced. Besides, in the same figure the experimental values (E_n/E_m) for nanocomposites epoxy polymer Na$^+$-montmorillonite (EP/MMT) at $T < T_g$ and $T > T_g$ (where T and T_g are testing and glass transition temperatures, respectively) are indicated by points. As one can see, for glassy epoxy matrix the experimental data correspond to the Eq. (16), that is, zero adhesional strength at a large friction coefficient and for devitrificated matrix to the Eq. (15), that is, the perfect adhesion between nanofiller and polymer matrix, described by Kerner equation. Let us note that the authors [17] explained the distinction indicated above by a much larger length of epoxy polymer segment in the second case.

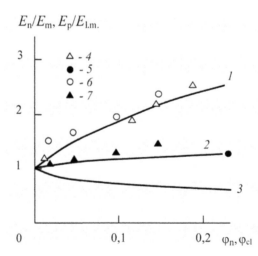

FIGURE 10.27   The dependences of reinforcement degree E_n/E_m and $E_p/E_{l.m.}$ on the contents of nanofiller φ_n and nanoclusters φ_{cl}, accordingly. 1÷3—theoretical dependences (E_n/E_m) (φ_n), corresponding to the Eqs. (15)÷(17); 4,5—the experimental data ($E_p/E_{l.m.}$) for Par at $T = T_g' ÷ T_g$ (4) and $T < T_g'$ (5); 6, 7—the experimental data (E_n/E_m) (φ_n) for EP/MMT at $T > T_g$ (6) and $T < T_g$ (7) [59].

To obtain the similar comparison for natural nanocomposite (polymer) is impossible, since at $T \geq T_g$ nanoclusters are disintegrated and polymer ceases to be quasi-two-phase system [5]. However, within the frameworks of two-stage glass transition concept [11] it has been shown, that at temperature T'_g, which is approximately equal to $T_g - 50$ K, instable (small) nanoclusters decay occurs, that results to loosely packed matrix devitrification at the indicated temperature [5]. Thus, within the range of temperature $T'_g \div T_g$ natural nanocomposite (polymer) is an analog of nanocomposite with glassy matrix [58]. As one can see, for the temperatures within the range of $T = T'_g \div T_g$ ($\varphi_{cl} = 0.06 \div 0.19$) the value $E_p/E_{l.m.}$ corresponds to the Eq. (15), that is, perfect adhesion nanoclusters-loosely packed matrix and at $T < T'_g$ ($\varphi_{cl} > 0.24$) – to the Eq. (16), that is, to zero adhesional strength at a large friction coefficient. Hence, the data of Fig. 10.27 demonstrated clearly the complete similarity, both qualitative and quantitative, of natural (PAr) and artificial (EP/MMT) nanocomposites reinforcement degree behavior. Another microcomposite model (e.g., accounting for the layered silicate particles strong anisotropy) application can change the picture quantitatively only. The data of Fig. 10.27 qualitatively give the correspondence of reinforcement degree of nanocomposites indicated classes at the identical initial conditions.

Hence, the analogy in behavior of reinforcement degree of polyarylate by nanoclusters and nanocomposite epoxy polymer/Na$^+$-montmorillonite by layered silicate gives another reason for the consideration of polymer as natural nanocomposite. Again strong influence of interfacial (intercomponent) adhesion level on nanocomposites of any class reinforcement degree is confirmed [17].

10.5 THE METHODS OF NATURAL NANOCOMPOSITES NANOSTRUCTURE REGULATION

As it has been noted above, at present it is generally acknowledged [2], that macromolecular formations and polymer systems are always natural nanostructural systems in virtue of their structure features. In this connection the question of using this feature for polymeric materials properties and operating characteristics improvement arises. It is obvious enough that for structure properties relationships receiving the quantitative nano-structural model of the indicated materials is necessary. It is also obvious that if the dependence of specific property on material structure state is unequivocal, then there will be quite sufficient modes to achieve this state. The cluster model of such state [3–5] is the most suitable for polymers amorphous state structure description. It has been shown, that this model basic structural element (cluster) is nanoparticles (nanocluster) (see Section 15.1). The cluster model was used successfully for cross-linked polymers structure and properties description [61]. Therefore the authors [62] fulfilled nanostructures regulation modes and have the latter influence on rarely cross-linked epoxy polymer properties study within the frameworks of the indicated model.

In Ref. [62] the studied object was an epoxy polymer on the basis of resin UP5–181, cured by iso-methyltetrahydrophthalic anhydride in the ratio by mass 1:0.56. Testing specimens were obtained by the hydrostatic extrusion method. The indicated method choice is due to the fact, that high hydrostatic pressure imposition in deformation process prevents the defects formation and growth, resulting to the material failure [64]. The extrusion strain ε_e was calculated and makes up 0.14, 0.25, 0.36, 0.43 and 0.52. The obtained by hydrostatic extrusion specimens were annealed at maximum temperature 353 K during 15 min.

The hydrostatic extrusion and subsequent annealing of rarely cross-linked epoxy polymer (REP) result to very essential changes of its mechanical behavior and properties, in addition unexpected ones enough. The qualitative changes of REP mechanical behavior can be monitored according to the corresponding changes of the stress–strain (σ–ε) diagrams, shown in Fig. 10.28. The initial REP shows the expected enough behavior and both its elasticity modulus E and yield stress σ_Y are typical for such polymers at testing temperature T being distant from glass transition temperature T_g on about 40 K [51]. The small ($\approx$ 3 MPa) stress drop beyond yield stress is observed, that is also typical for amorphous polymers [61]. However, REP extrusion up to ε_e = 0.52 results to stress drop $\Delta\sigma_Y$ ("yield tooth") disappearance and to the essential E and σ_Y reduction. Besides, the diagram σ–ε itself is now more like the similar diagram for rubber, than for glassy polymer. This specimen annealing at maximum temperature T_{an} = 353 K gives no less strong, but diametrically opposite effect – yield stress and elasticity modulus increase sharply (the latter in about twice in comparison with the initial REP and more than one order in comparison with the extruded specimen). Besides, the strongly pronounced "yield tooth" appears. Let us note, that specimen shrinkage at annealing is small ($\approx$10%), that makes up about 20% of ε_e [62].

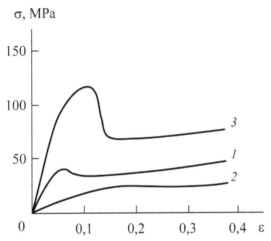

FIGURE 10.28   The stress–strain (σ–ε) diagrams for initial (1), extruded up to $\varepsilon_e = 0.52$ (2) and annealed (3) REP samples [62].

The common picture of parameters E and σ_Y change as a function of ε_e is presented in Figs. 10.29 and 10.30 accordingly. As one can see, both indicated parameters showed common tendencies at ε_e change: up to $\varepsilon_e \approx$ 0.36 inclusive E and σ_Y weak increase at ε_e growth is observed, moreover their absolute values for extruded and annealed specimens are close, but at $\varepsilon_e >$ 0.36 the strongly pronounced antibatness of these parameters for the indicated specimen types is displayed. The cluster model of polymers amorphous state structure and developed within its frameworks polymers yielding treatment allows to explain such behavior of the studied samples [35, 65].

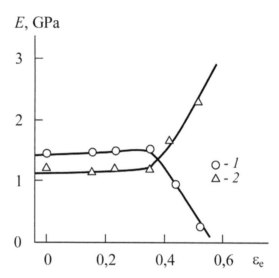

FIGURE 10.29 The dependences of elasticity modulus E_p on extrusion strain ε_e for extrudated (1) and annealed (2) REP [62].

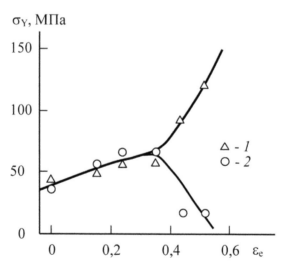

FIGURE 10.30 The dependences of yield stress σ_Y on extrusion strain ε_e for extrudated (1) and annealed (2) REP [62].

The cluster model supposes that polymers amorphous state structure represents the local order domains (nanoclusters), surrounded by loosely packed matrix. Nanoclusters consist of several collinear densely packed statistical segments of different macromolecules and in virtue of this they offer the analog of crystallite with stretched chains. There are two types of nanoclusters—*stable*, consisting of a relatively large segments number, and *instable*, consisting of a less number of such segments [65]. At temperature increase or mechanical stress application the instable nanoclusters disintegrate in the first place, that results to the two well-known effects. The first from them is known as two-stage glass transition process [11] and it supposes that at $T'_g = T_g - 50$ K disintegration of instable nanoclusters, restraining loosely packed matrix in glass state, occurs that defines devitrification of the latter [3, 5]. The well-known rapid polymers mechanical properties reduction at approaching to T_g [51] is the consequence of this. The second effect consists of instable nanoclusters decay at σ_Y under mechanical stress action, loosely packed matrix mechanical devitrification and, as consequence, glassy polymers rubber-like behavior on cold flow plateau [65]. The stress drop $\Delta\sigma_Y$ beyond yield stress is due to just instable nanoclusters decay and therefore $\Delta\sigma_Y$ value serves as characteristic of these nanoclusters fraction [5]. Proceeding from this brief description, the experimental results, adduced in Figs. 10.28 ÷ 10.30, can be interpreted.

The rarely cross-linked epoxy polymer on the basis of resin UP5–181 has low glass transition temperature T_g, which can be estimated according to shrinkage measurements data as equal ≈ 333K. This means, that the testing temperature $T = 293$ K and T'_g for it are close, that is confirmed by small $\Delta\sigma_Y$ value for the initial REP. It assumes nanocluster (nanostructures) small relative fraction φ_{cl} [3–5] and, since these nanoclusters have arbitrary orientation, ε_e increase results rapidly enough to their decay, that induces loosely packed matrix mechanical devitrification at $\varepsilon_e > 0.36$. Devitrificated loosely packed matrix gives insignificant contribution to E_p [66, 67], equal practically to zero, that results to sharp (discrete) elasticity modulus decrease. Besides, at $T > T'_g$ φ_{cl} rapid decay is observed, that is, segments number decrease in both stable and instable nanocluster [5]. Since just these parameters (E and φ_{cl}) check σ_Y value, then their decrease defines yield stress sharp lessening. Now extruded at $\varepsilon_e > 0.36$ REP presents as matter of fact rubber with high cross-linking degree, which is reflected by its diagram σ–ε (Fig. 10.28, curve 2).

The polymer oriented chains shrinkage occurs at the extruded REP annealing at temperature higher than T_g. Since this process is realized within a narrow temperature range and during a small time interval, then a large number of instable nanoclusters is formed. This effect is intensified by available molecular orientation, that is, by preliminary favorable segments arrangement, and it is reflected by $\Delta\sigma_Y$ strong increase (Fig. 10.28, curve 3).

The φ_{cl} enhancement results to E_p growth (Fig. 10.29) and φ_{cl} and E_p combined increase–to σ_Y considerable growth (Fig. 10.30).

The considered structural changes can be described quantitatively within the frameworks of the cluster model. The nanoclusters relative fraction φ_{cl} can be calculated according to the method, stated in Ref. [68].

The shown in Fig. 10.31 dependences $\varphi_{cl}(\varepsilon_e)$ have the character expected from the adduced above description and are its quantitative conformation. The adduced in Fig. 10.32 dependence of density ρ of REP extruded specimens on ε_e is similar to the dependence $\varphi_{cl}(\varepsilon_e)$, that was to be expected, since densely packed segments fraction decrease must be reflected in ρ reduction.

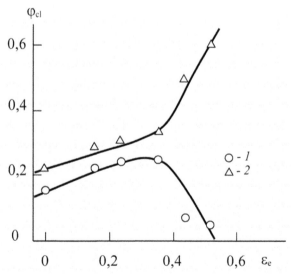

FIGURE 10.31 The dependences of nanoclusters relative fraction φ_{cl} on extrusion strain ε_e for extruded (1) and annealed (2) REP [62].

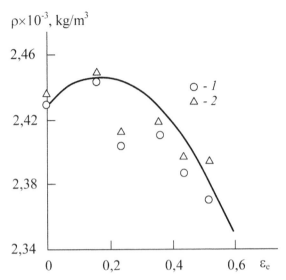

FIGURE 10.32   The dependence of specimens density ρ on extrusion strain ε_e for extruded (1) and annealed (2) REP [62].

In Ref. [69] the supposition was made that ρ change can be conditioned to microcracks network formation in specimen that results to ρ reduction at large ε_e (0.43 and 0.52), which are close to the limiting ones. The ρ relative change ($\Delta\rho$) can be estimated according to the equation

$$\Delta\rho = \frac{\rho^{max} - \rho^{min}}{\rho^{max}},\tag{33}$$

where ρ^{max} and ρ^{min} are the greatest and the smallest density values. This estimation gives $\Delta\rho \approx 0.01$. This value can be reasonable for free volume increase, which is necessary for loosely matrix devitrification (accounting for closeness of T and T'_g), but it is obviously small if to assume as real microcracks formation. As the experiments have shown, REP extrusion at $\varepsilon_e > 0.52$ is impossible owing to specimen cracking during extrusion process. This allows to suppose that value $\varepsilon_e = 0.52$ is close to the critical one. Therefore the critical dilatation $\Delta\delta_{cr}$ value, which is necessary for microcracks cluster formation, can be estimated as follows [40]:

$$\Delta\delta_{cr} = \frac{2(1+\nu)(2-3\nu)}{11-19\nu},\tag{34}$$

where ν is Poisson's ratio.

Accepting the average value $\nu \approx 0.35$, we obtain $\Delta\delta_{cr} = 0.60$, that is essentially higher than the estimation $\Delta\rho$ made earlier. These calculations assume that ρ decrease at $\varepsilon_e = 0.43$ and 0.52 is due to instable nanoclusters decay and to corresponding REP structure loosening.

The stated above data give a clear example of large possibilities of polymer properties operation through its structure change. From the plots of Fig. 10.29 it follows that annealing of REP extruded up to $\varepsilon_e = 0.52$ results to elasticity modulus increase in more than eight times and from the data of Fig. 10.30 yield stress increase in six times follows. From the practical point of view the extrusion and subsequent annealing of rarely cross-linked epoxy polymers allow to obtain materials, which are just as good by stiffness and strength as densely cross-linked epoxy polymers, but exceeding the latter by plasticity degree. Let us note, that besides extrusion and annealing other modes of polymers nanostructure operation exist: plasticization [70], filling [26, 71], films obtaining from different solvents [72] and so on.

Hence, the stated above results demonstrated that neither cross-linking degree nor molecular orientation level defined cross-linked polymers final properties. The factor, controlling properties is a state of suprasegmental (nanocluster) structure, which, in its turn, can be goal-directly regulated by molecular orientation and thermal treatment application [62].

In the stated above treatment not only nanostructure integral characteristics (macromolecular entanglements cluster network density ν_{cl} or nanocluster relative fraction φ_{cl}), but also separate nanocluster parameters are important (see section 1). In this case of particulate-filled polymer nanocomposites (artificial nanocomposites) it is well-known, that their elasticity modulus sharply increases at nanofiller particles size decrease [17]. The similar effect was noted above for REP, subjected to different kinds of processing (*see* Fig. 10.28). Therefore the authors [73] carried out the study of the dependence of elasticity modulus E on nanoclusters size for REP.

It has been shown earlier on the example of PC, that the value E_p is defined completely by natural nanocomposite (polymer) structure according to the Eq. (32) (*see* Fig. 10.26).

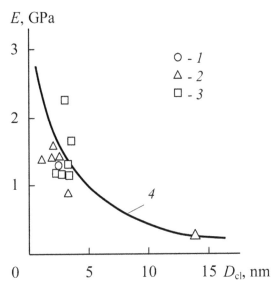

FIGURE 10.33 The dependence of elasticity modulus E_p on nanoclusters diameter D_{cl} for initial (1), extruded (2) and annealed (3) REP. 4—calculation according to the Eq. (32) [73].

In Fig. 10.33, the dependence of E_p on nanoclusters diameter D_{cl}, determined according to the equation (in Chapter 9), for REP subjected to the indicated processing kinds at ε_e values within the range of 0.16 0.52 is adduced. As one can see, like in the case of artificial nanocomposites, for REP strong (approximately of order of magnitude) growth is observed at nanoclusters size decrease from 3 up to 0.9 nm. This fact confirms again, that REP elasticity modulus is defined by neither cross-linking degree nor molecular orientation level, but it depends only on epoxy polymer nanocluster structure state, simulated as natural nanocomposite [73].

Another method of the theoretical dependence $E_p(D_{cl})$ calculation for natural nanocomposites (polymers) is given in Ref. [74]. The authors [75] have shown, that the elasticity modulus E value for fractal objects, which are polymers [4], is given by the following percolation relationship:

$$K_T, G \sim (p-p_c)^{\eta}, \tag{35}$$

where K_T is bulk modulus, G is shear modulus, p is solid-state component volume fraction, p_c is percolation threshold, η is exponent.

The following equation for the exponent η was obtained at a fractal structure simulation as Serpinsky carpet [75]:

$$\frac{\eta}{v_p} = d - 1, \tag{36}$$

where v_p is correlation length index in percolation theory, d is dimension of Euclidean space, in which a fractal is considered.

As it is known [4], the polymers nanocluster structure represents itself the percolation system, for which $p = \varphi_{cl}$, $p_c = 0.34$ [35] and further it can be written:

$$\frac{R_{cl}}{l_{st}} \sim \left(\phi_{cl} - 0,34\right)^{i_p}, \tag{37}$$

where R_{cl} is the distance between nanoclusters, determined according to the Eq. (4.63), l_{st} is statistical segment length, v_p is correlation length index, accepted equal to 0.8 [77].

Since in the considered case the change E_p at n_{cl} variation is interesting first of all, then the authors [74] accepted $v_{cl} = \text{const} = 2.5 \times 10^{27}$ m^{-3}, $l_{st} = \text{const} = 0.434$ nm. The value E_p calculation according to the Eqs. (35) and (37) allows to determine this parameter according to the formula [74]:

$$E_p = 28,9\left(\phi_{cl} - 0,34\right)^{(d-1) \, p}, \text{ GPa}. \tag{38}$$

In Fig. 10.34, the theoretical dependence (a solid line) of E_p on nanoclusters size (diameter) D_{cl}, calculated according to the Eq. (38) is adduced. As one can see, the strong growth E_p at D_{cl} decreasing is observed, which is identical to the shown one in Fig. 10.33. The adduced in Fig. 10.34 experimental data for REP, subjected to hydrostatic extrusion and subsequent annealing, correspond well enough to calculation according to the equation (38). The decrease D_{cl} from 3,2 up to 0,7 nm results again to E_p growth on order of magnitude [74].

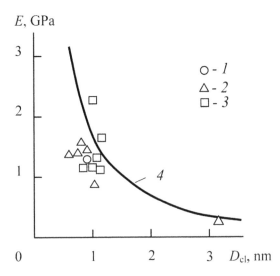

FIGURE 10.34 The dependence of elasticity modulus E_p on nanoclusters diameter D_{cl} for initial (1), extruded (2) and annealed (3) REP. 4—calculation according to the Eq. (38) [74].

The similar effect can be obtained for linear amorphous polycarbonate (PC) as well. Calculation according to the Eq. (38) shows, n_{cl} reduction from 16 (the experimental value n_{cl} at $T = 293$ K for PC [5]) up to two results to E_p growth from 1.5 up to 5.8 GPa and making of structureless ($n_{cl} = 1$) PC will allow to obtain $E_p \approx 9.2$ Gpa, that is, comparable with obtained one for composites on the basis of PC.

Hence, the stated in the present chapter results give purely practical aspect of such theoretical concepts as the cluster model of polymers amorphous state stricture and fractal analysis application for the description of structure and properties of polymers, treated as natural nanocomposites. The necessary nanostructure goal-directed making will allow to obtain polymers, not yielding (and even exceeding) by their properties to the composites, produced on their basis. Structureless (defect-free) polymers are imagined the most perspective in this respect. Such polymers can be natural replacement for a large number of elaborated at present polymer nanocomposites. The application of structureless polymers as artificial nanocomposites polymer matrix can give much larger effect. Such approach allows to obtain polymeric materials, comparable by their characteristics with metals (e.g., with aluminum).

KEYWORDS

- fractal analysis
- natural composites
- polymers
- properties
- structure
- theoretical model

REFERENCES

1. Kardos, I. L., Raisoni, I. The Potential Mechanical Response of Macromolecular Systems-F Composite Analogy. Polymer Eng. Sci., 1975, v.15, N3, p.183–189.
2. Ivanches, S. S., Ozerin, A. N. A Nanostructures In Polymeric Systems.Vysokomolek Soed.B, 2006, v.48, N8, p.1531–1544.
3. Kozlov, G. V., Novikov, V. U. The Cluster Model of Polymers Amorphous State. Uspekhi Fizicheskikh Nauk, 2001, v.171, N7, p. 717–764.
4. Kozlov, G. V., Zaikov, G. E. Structure of the Polymer Amorphous State. Utrecht, Boston, Brill Academic Publishers, 2004, 465 p.
5. Kozlov, G. V., Ovcharenko, E. N., Mikitaev, A. K. Structure of the Polymer Amorphous State. Moscow, Publishers of the, D. I.Mendeleev RKhTU, 2013, 392 p.
6. Kozlov, G. V., Novikov, V. U. Synergetics and Fractal Analysis of Cross-Linked Polymers. Moscow, Klassika, 2013, 112 p.
7. Burya, A. I., Kozlov, G. V., Novikov, V. U., Ivanova, V. S. Synergetics of Supersegmental Structure of Amorphous Glassy Polymers. Mater. of 3-rd Intern.Conf."Research and Development in Mechanical Industry-RaDMI-03", September 19–23, 2003, Herceg Novi, Serbia and Montenegro, p. 645–647.
8. Bashorov, M. T., Kozlov, G. V., Mikitaev, A. K. Nanostructures and Properties of Amorphous Glassy Polymers. Moscow, Publishers of the, D. I. Mendeleev RKhTU, 2010, 269 p.
9. Malamatov, A.Kh., Kozlov, G. V., Mikitaev, M. A. Reinforcement Mechanisms of Polymer Nanocomposites. Moscow, Publishers of the, D. I. Mendeleev RKhTU, 2006, 240 p.
10. Kozlov, G. V., Gazaev, M. A., Novikov, V. U., Mikitaev, A. K. Simulation of Amorphous Polymers Structure as Percolation Cluster. Pis'ma v ZhTF, 1996, v.22, N16, p. 31–38.
11. Belousov, V. N., Kotsev, B.Kh., Mikitaev, A. K. Two-Step of Amorphous Polymers Glass Transition Doklady ANSSSR, 1983, v.270, N5, p. 1145–1147.
12. Ivanova, V. S., Kuzeev, I. R., Zakirnichnaya, M. M. Synergetics and Fractals. Universality of Metal Mechanical Behaviour. Ufa, Publishers of UGNTU, 1998, 366p.

13. Berstein, V. A., Egorov, V. M. Differential scanning calorimetry in Physics-Chemistry of the Polymers. Leningrad, Khimiya, 1990, 256p.
14. Bashorov, M. T., Kozlov, G. V., Mikitaev, A. K. A Nanoclusters Synergetics in Amorphous Glassy Polymers Structure. Inzhenernaya Fizika, 2009, N4, p.39–42.
15. Bashorov, M. T., Kozlov, G. V., Mikitaev, A. K. A Nanostructures in Polymers: Formation synergetics, Regulation Methods and Influence on the properties. Materialovedenie, 2009, N9, p. 39–51.
16. Shevchenko, V.Ya., Bal'makov, M. D. A Particles-Centravs as Nanoworld objects. Fizika I Khimiya Stekla, 2002, v.28, N6, p.631–636
17. Mikitaev, A. K., Kozlov, G. V., Zaikov, G. E. Polymer Nanocomposites: Variety of Structural Forms and Applications. New York, Nova Science Publishers, Inc., 2008, 318p.
18. Buchachenko, A. L. The Nanochemistry Direct Way to High Technologies of New Century. Uspekhi Khimii, 2003, v.72, N5, p.419–437
19. Formanis, G. E. Self-Assembly of Nanoparticles is Nanoworld Special Properties Spite. Proceedings of Intern. Interdisciplinary Symposium "Fractals and Applied Synergetics", FiPS-03″, Moscow, Publishers of MGOU, 2003, 303–308.
20. Bashorov, M. T., Kozlov, G. V., Shustov, G. B., Mikitaev, A. K. The Estimation of Fractal Dimension of Nanoclusters Surface in Polymers. Izvestiya Vuzov, Severo-Kavkazsk. region, estestv. nauki, 2009, N6, p.44–46.
21. Magomedov, G. M., Kozlov, G. V. Synthesis, Structure and Properties of Cross-Linked Polymers and Nanocomposites on its Basis. Moscow, Publishers of Natural Sciences Academy, 2010, 464p.
22. Kozlov, G. V. Polymers as Natural Nanocomposites: the Missing Opportunities. Recent Patents on Chemical Engineering, 2011, v.4, N1, p.53–77.
23. Bovenko, V. N., Startsev, V. M. The Discretely Wave Nature of Amorphous Poliimide Supramolecular Organization. Vysokomolek. Soed. B, 1994, v.36, N6, p.1004–1008
24. Bashorov, M. T., Kozlov, G. V., Mikitaev, A. K. Polymers as Natural Nanocomposites: An Interfacial Regions Identification. Proceedings of 12th Intern. Symposium "Order, Disorder and Oxides Properties." Rostov-na-Donu-Loo, September 17–22, 2009, p.280–282.
25. Magomedov, G. M., Kozlov, G. V., Amirshikhova, Z. M. Cross-Linked Polymers as Natural Nanocomposites: An Interfacial Region Identification. Izvestiya DGPU, estestv. I tochn. nauki, 2013, N4, p.19–22
26. Kozlov, G. V., Yanovskii Yu.G., Zaikov, G. E. Structure and Properties of Particulate-Filled Polymer Composites: the Fractal Analysis. New York, Nova Science Publishers, Inc., 2010, 282p.
27. Bashorov, M. T., Kozlov, G. V., Shustov, G. B., Mikitaev, A. K. Polymers as Natural Nanocomposites: the Filling Degree Estimations. Fundamental'nye Issledovaniya, 2009, N4, p.15–18.
28. Vasserman, A. M., Kovarskii, A. L. A Spin Probes and Labels in Physics-Chemistry of Polymers. Moscow, Nauk, 2013, 246p.
29. Korst, N. N., Antsiferova, L. I. A Slow Molecular Motions Study by Stable Radicals EPR Method. Uspekhi Fizicheskikh Nauk, 1978, v.126, N1, p.67–99.
30. Yech, G. S. The General Notions on Amorphous Polymers Structure. Local Order and Chain Conformation Degrees. Vysokomolek.Soed.A, 2013, v.21, N11, p.2433–2446.

31. Perepechko, I. I. Introduction in Physics of Polymers. Moscow, Khimiya, 1978, 312p.
32. Kozlov, G. V., Zaikov, G. E. The Generalized Description of Local Order in Polymers. In book: Fractals and Local Order in Polymeric Materials. Ed.Kozlov, G. V., Zaikov, G. E. New York, Nova Science Publishers, Inc., 2001, p.55–63.
33. Tager, A. A. Physics-Chemistry of Polymers. Moscow, Khimiya, 1978, 416p.
34. Bashorov, M. T., Kozlov, G. V., Malamatov, A.Kh., Mikitaev, A. K. Amorphous Glassy Polymers Reinforcement Mechanisms by Nanostructures. Mater. of IV Intern.Sci.-Pract.Conf. "New Polymer Composite Materials." Nal'chik, KBSU, 2008, p.47–51.
35. Bobryshev, A. N., Koromazov, V. N., Babin, L. O., Solomatov, V. I. Synergetics of Composite Materials. Lipetsk, NPO ORIUS, 1994, 154p.
36. Aphashagova, Z.Kh., Kozlov, G. V., Burya, A. T., Mikitaev, A. K. The Prediction of particulate-Filled Polymer Nanocomposites Reinforcement Degree. materialovedenie, 2007, N9, p.10–13.
37. Sheng, N., Boyce, M. C., Parks, D. M., Rutledge, G. C., Ales, J. I., Cohen, R. E. Multiscale Micromechanical modeling of Polymer/Clay Nanocomposites and the Effective Clay Particle. Polymer, 2004, v.45, N2, p.487–506.
38. Dickie, R. A. The Mechanical Properties (Small Strains) of Multiphase Polymer Blends. In book: Polymer Blends. Ed. Paul, D. R., Newman, S. New York, San Francisko, London, Academic Press, 1980, v.1, p.397–437.
39. Ahmed, S., Jones, F. R. A review of particulate Reinforcement Theories for Polymer Composites. J.Mater.Sci., 1990, v.25, N12, p.4933–4942.
40. Balankin, A. S. Synergetics of Deformable Body. Moscow, Publishers of Ministry Defence SSSR, 2013, 404p.
41. Bashorov, M. T., Kozlov, G. V., Mikitaev, A. K. Polymers as Natural Nanocomposites: Description of Elasticity Modulus within the Frameworks of Micromechanical Models. Plast. Massy, 2010, N11, p.41–43
42. Lipatov Yu.S. Interfacial Phenomena in Polymers. Kiev, Naukova Dumka, 1980, 260p.
43. Yanovskii Yu.G., Bashorov, M. T., Kozlov, G. V., Karnet Yu.N. Polymeric Mediums as Natural Nanocomposites: Intercomponont Interactions Geometry. Proceedings of All-Russian Conf. "Mechanics and Nanomechanics of Structurally Complex and Heterogeneous Mediums Achievements, Problems, Perspectives." Moscow, IPROM, 2012, p.110–117.
44. Tugov, I. I., Shaulov, A.Yu. A Particulate-Filled Composites Elasticity Modulus. Vysokomolek. Soed. B, 1990, v.32, N7, p.527–529.
45. Piggott, M. R., Leidner, Y. Microconceptions about Filled Polymers. Y.Appl.Polymer Sci., 1974, v.18, N7, p.1619–1623.
46. Chen, Z.-Y., Deutch, Y. M., Meakin, P. Translational Friction Coefficient of Diffusion Limited Aggregates. Y.Chem. Phys., 1984, v.80, N6, p.2982–2983.
47. Kozlov, G. V., Beloshenko, V. A., Varyukhin, V. N. Simulation of Cross-Linked Polymers Structure as Diffusion-Limited Aggregate. Ukrainskii Fizicheskii Zhurnal, 1998, v.43, N3, p.322–323.
48. Novikov, V. U., Kozlov, G. V., Burlyan, O. Y. The Fractal Approach to Interfacial Layer in Filled Polymers. Mekhanika Kompozitnykh Materialov, 2013, v.36, N1, p.3–32.
49. Stanley, E. H. A Fractal Surfaces and "Termite" Model for Two-Component Random Materials. In book: Fractals in Physics. Ed. Pietronero, L., Tosatti, E. Amsterdam, Oxford, New York, Tokyo, North-Holland, 1986, p.463–477.

50. Bashorov, M. T., Kozlov, G. V., Zaikov, G. E., Mikitaev, A. K. Polymers as Natural Nanocomposites: Adhesion between Structural Components. Khimicheskaya Fizika i Mezoskopiya, 2013, v.11, N2, p.196–203.
51. Dibenedetto, A. T., Trachte, K. L. The Brittle Fracture of Amorphous Thermoplastic Polymers. Y.Appl. Polymer Sci., 1970, v.14, N11, p.2249–2262.
52. Burya, A. I., Lipatov Yu.S., Arlamova, N. T., Kozlov, G. V. Patent by Useful Model N27 199. Polymer composition. It is registered in Ukraine Patents State Resister October 25–2007.
53. Novikov, V. U., Kozlov, G. V. Fractal Parametrization of Filled Polymers structure. Mekhanika Kompozitnykh Materialov, 1999, v.35, N3, p.269–290.
54. Potapov, A. A. A Nanosystems Design Principles. Nano- i Mikrosistemnaya Tekhnika, 2008, v.3, N4, p.277–280.
55. Bashorov, M. T., Kozlov, G. V., Zaikov, G. E., Mikitaev, A. K. Polymers as Natural nanocomposites. 3. The Geometry of Intercomponent Interactions. Chemistry and Chemical Technology, 2009, v.3, N4, p.277–280.
56. Bashorov, M. T., Kozlov, G. V., Zaikov, G. E., Mikitaev, A. K. Polymers as Natural Nanocomposites. 1. The Reinforcement Structural Model. Chemistry and Chemical Technology, 2009, v.3, N2, p.107–110.
57. Edwards, D. C. Polymer-Filler Interactions in Rubber Reinforcement. I. Mater. Sci., 1990, v.25, N12, p.4175–4185.
58. Bashorov, M. T., Kozlov, G. V., Mikitaev, A. K. Polymers as Natural nanocomposites: the Comparative Analysis of Reinforcement Mechanism. Nanotekhnika, 2009, N4, p.43–45.
59. Bashorov, M. T., Kozlov, G. V., Zaikov, G. E., Mikitaev, A. K. Polymers as Natural Nanocomposites. 2. The Comparative Analysis of Reinforcement Mechanism. Chemistry and Chemical Technology, 2009, v.3, N3, p.183–185.
60. Chen, Y.-S., Poliks, M. D., Ober, C. K., Zhang, Y., Wiesner, U., Giannelis, E. Study of the Interlayer Expansion Mechanism and Thermal-Mechanical Properties of Surface-Initiated Epoxy Nanocomposites. Polymer, 2002, v.43, N17, p.4895–4904.
61. Kozlov, G. V., Beloshenko, V. A., Varyukhin, V. N., Lipatov Yu.S. Application of Cluster Model for the Description of Epoxy Polymers Structure and Properties. Polymer, 1999, v.40, N4, p.1045–1051.
62. Bashorov, M. T., Kozlov, G. V., Mikitaev, A. K. Nanostructures in Cross-Linked Epoxy Polymers and their Influence on Mechanical Properties. Fizika I Khimiya Obrabotki Materialov, 2013, N2, p.76–80.
63. Beloshenko, V. A., Shustov, G. B., Slobodina, V. G., Kozlov, G. V., Varyukhin, V. N., Temiraev, K. B., Gazaev, M. A. Patent on Invention "The Method of Rod-Like Articles Manufacture from Polymers." Clain for Invention Rights N95109832. Patent N2105670. Priority: 13 June 1995. It is Registered in Inventions State Register of Russian Federation February 27–1998.
64. Aloev, V. Z., Kozlov, G. V. Physics of Orientational Phenomena in Polymeric Materials. Nalchik, Polygraph-service and T, 2002, 288p.
65. Kozlov, G. V., Beloshenko, V. A., Garaev, M. A., Novikov, V. U. Mechanisms of Yielding and Forced High-Elasticity of Cross-Linked Polymers. Mekhanika Kompozitnykh Materialov, 2013, v.32, N2, p.270–278.

66. Shogenov, V. N., Belousov, V. N., Potapov, V. V., Kozlov, G. V., Prut, E. V. The Glassy Polyarylatesurfone Curves Stress-Strain Description within the Frameworks of High-Elasticity Concepts. Vysokomolek.Soed.F, 1991, v.33, N1, p.155–160.
67. Kozlov, G. V., Beloshenko, V. A., Shogenov, V. N. The Amorphous Polymers Structural Relaxation Description within the Frameworks of the Cluster Model. Fiziko-Khimicheskaya Mekhanika Materialov, 2013, v.35, N5, p.105–108.
68. Kozlov, G. V., Burya, A. I., Shustov, G. B. The Influence of Rotating Electromagnetic Field on Glass Transition and Structure of Carbon Plastics on the Basis of lhenylone. Fizika I Khimiya Obrabotki Materialov, 2005, N5, p.81–84.
69. Pakter, M. K., Beloshenko, V. A., Beresnev, B. I., Zaika, T. R., Abdrakhmanova, L. A., Berai, N. I. Influence of Hydrostatic Processing on Densely Cross-Linked Epoxy Polymers Structural Organization Formation. Vysokomolek.Soed.F, 2013, v.32, N10, p.2039–2046.
70. Kozlov, G. V., Sanditov, D. S., Lipatov Yu.S. Structural and Mechanical Properties of Amorphous Polymers in Yielding Region. In book: Fractals and Local Order in Polymeric Materials. Ed. Kozlov, G. V., Zaikov, G. E. New York, Nova Science Publoshers, Inc., 2001, p.65–82.
71. Kozlov, G. V., Yanovskii Yu.G., Zaikov, G. E. Synergetics and Fractal Analysis of Polymer Composites Filled with Short Fibers. New York, Nova Science Publishers, Inc., 2011, 223p.
72. Shogenov, V. N., Kozlov, G. V. Fractal Clusters in Physics-Chemistry of Polymers. Nal'chik, Polygraphservice and T, 2002, 270p.
73. Kozlov, G. V., Mikitaev, A. K. Polymers as Natural Nanocomposites: Unrealized Potential. Saarbrücken, Lambert Academic Publishing, 2010, 323p.
74. Magomedov, G. M., Kozlov, G. V., Zaikov, G. E. Structure and Properties of Cross-Linked Polymers. Shawbury, A Smithers Group Company, 2011, 492p.
75. Bergman, D. Y., Kantor, Y. Critical Properties of an Elastic Fractal. Phys. Rev. Lett., 1984, v.53, N6, p.511–514.
76. Malamatov, A. Kh., Kozlov, G. V. The Fractal Model of Polymer-Polymeric Nanocomposites Elasticity. Proceedings of Fourth Intern. Interdisciplinary Symposium "Fractals and Applied Synergetics FaAS-05." Moscow, Interkontakt Nauka, 2005, p.119–122.
77. Sokolov, I. M. Dimensions and Other Geometrical Critical Exponents in Percolation Theory. Uspekhi Fizicheskikh Nauk, 2013, v.151, N2, p.221–248.

CHAPTER 11

A LECTURE NOTE ON CLUSTER MODEL OF POLYMERS AMORPHOUS STATE STRUCTURE

G. V. KOZLOV, I. V. DOLBIN, JOZEF RICHERT, O. V. STOYANOV, and G. E. ZAIKOV

CONTENTS

11.1   INTRODUCTION

A cluster model of polymers amorphous state structure allows to introduce principally new treatment of structure defect (in the full sence of
this term) for the indicated state [1, 2]. As it is known [3], real solids
structure contains a considerable number of defects. The given concept
is the basis of dislocations theory, widely applied for crystalline solids
behavior description. Achieved in this field successes predetermine the attempts of authors number [4–11] to use the indicated concept in reference
to amorphous polymers. Additionally, used for crystalline lattices notions
are often transposed to the structure of amorphous polymers. As a rule,
the basis for this transposition serves formal resemblance of stress–strain
(σ–ε) curves for crystalline and amorphous solids.

In relation to the structure of amorphous polymers for a long time the
most ambiguous point [12–14] was the presence or the absence of the local (short-range) order in this connection points of view of various authors
on this problem were significantly different. The availability of the local
order can significantly affect the definition of the structure defect in amorphous polymers, if in the general case the order-disorder transition or vice
versa is taken for the defect. For example, any violation (interruption) of
the long-range order in crystalline solids represents a defect (dislocation,
vacancy, etc.), and a monocrystal with the perfect long-range order is the
ideal defect-free structure with the perfect long-range order. It is known
[15], that sufficiently bulky samples of 100% crystalline polymer cannot
be obtained, and all characteristics of such hypothetical polymers are determined by the extrapolation method. That is why the Flory "felt" model [16,
17] can be suggested as the ideal defect-free structure for amorphous state
of polymers. This model assumes that amorphous polymers consist of in-
terpenetrating macromolecular coils personifying the full disorder (chaos).
Proceeding from this, as the defect in polymers amorphous state a violation (interruption) of full disorder must be accepted, that is, formation of
the local (or long-range) order [1]. It should also be noted that the formal
resemblance of the curves σ–ε for crystalline solids and amorphous polymers appears far incomplete and the behavior of these classes of materials
displays principal differences, which will be discussed in detail below.

Turning back to the suggested concept of amorphous polymer structural defect, let us note that a segment including in the cluster can be

considered as the linear defect – the analog of dislocation in crystalline solids. Since in the cluster model the length of such segment is accepted equal to the length of statistical segment, l_{st}, and their amount per volume unit is equal to the density of entanglements cluster network, v_{cl}, then the density of linear defects, ρ_d, per volume unit of the polymer can be expressed as follows [1]:

$$\rho_d = v_{cl} \times l_{st}. \tag{1}$$

The offered treatment allows application of well-developed mathematical apparatus of the dislocation theory for the description of amorphous polymers properties. Its confirmation by the X-raying methods was stated in Ref. [18].

Further on, the rightfulness of application of the structural defect concept to polymers yielding process description will be considered. As a rule, previously assumed concepts of defects in polymers were primarily used for the description of this process or even exclusively for this purpose [4–11]. Theoretical shear strength of crystals was first calculated by Frenkel, basing on a simple model of two atoms series, displaced in relation to one another by the shear stress (Fig. 11.1a) [3]. According to this model, critical shear stress τ_0 is expressed as follows [3]:

$$\tau_0 = \frac{G}{2\pi}, \tag{2}$$

where G is the shear modulus.

Slightly changed, this model was used in the case of polymers yielding [6], wherefrom the following equation was obtained:

$$\tau_{0Y} = \frac{G}{\pi\sqrt{3}}, \tag{3}$$

where τ_{0Y} is a theoretical value of the shear stress at yielding.

Special attention should be paid to the fact that characterizes principally different behavior of crystalline metals compared with polymers. As it is known [3, 19], τ_{0Y}/τ_Y ratio (where τ_Y is experimentally determined shear stress at yielding) is much higher for metals than for polymers. For five metals possessing the face-centered cubic or hexagonal lattices the following ratios were obtained: $\tau_{0Y}/\tau_Y = 37{,}400 \div 22{,}720$ (according to the data of Ref [3]), whereas for five polymers this ratio makes $2.9 \div 6.3$ [6]. In essence, sufficient closeness of τ_{0Y}/τ_Y ratio values to one may already be the proof for the possibility of realizing of Frenkel mechanism in polymers (in contrast with metals), but it will be shown below that for polymers a small modification of the law of shear stress τ periodic change used commonly gives τ_{0Y}/τ_Y values very close to one [20].

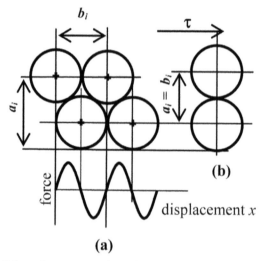

FIGURE 11.1 Schematic representation of deformation of two sequences of atoms according to the Frenkel model. Positions before (a) and after (b) deformation [3].

As it has been shown in Ref. [21], dislocation analogies are also true for amorphous metals. In essence, the authors [21] consider the atoms construction distortion (which induces appearance of elastic stress fields) as a linear defect (dislocation) being practically immovable. It is clear that such approach correlates completely with the offered above structural defect concept. Within the frameworks of this concept, Fig. 11.1a may be considered as a cross-section of a cluster (crystallite) and, hence, the shear of segments in the latter according to the Frenkel mechanism—as a mechanism limiting yielding process in polymers. This is proved by the experimental data [32], which shown that glassy polymers yielding process is realized namely in densely packed regions. Other data [23] indicate that these densely packed regions are clusters. In other words, one can state that yielding process is associated with clusters (crystallites) stability loss in the shear stresses field [24].

In Ref. [25] the asymmetrical periodic function is adduced, showing the dependence of shear stress τ on shear strain γ_{sh} (Fig. 11.2). As it has been shown before [19], asymmetry of this function and corresponding decrease of the energetic barrier height overcome by macromolecules segments in the elementary yielding act are due to the formation of fluctuation free volume voids during deformation (that is the specific feature of polymers [26]). The data in Fig. 11.2 indicate that in the initial part of periodic curve from zero up to the maximum dependence of τ on displacement x can be simulated by a sine-shaped function with a period shorter than in Fig. 11.1. In this case, the function $\tau(x)$ can be presented as follows:

$$\tau(x) = k\sin\left(\frac{6\pi x}{b_i}\right), \tag{4}$$

that is fully corresponds to Frenkel conclusion, except for arbitrarily chosen numerical coefficient in brackets (6 instead of 2).

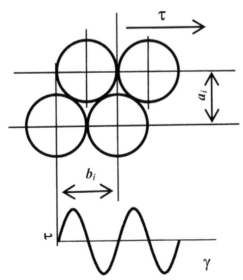

FIGURE 11.2 Schematic picture of shear deformation and corresponding stress–strain (τ–γ_{sh}) function [25].

Further calculation of τ_o by method, described in Ref [3], and its comparison with the experimental values τ_x indicated their close correspondence for nine amorphous and semicrystalline polymers (Fig. 11.3), which proves the possibility of realization of the above-offered yielding mechanism at the segmental level [18].

Inconsistency of τ_{oY} and τ_Y values for metals results to a search for another mechanism of yielding realization. At present, it is commonly accepted that this mechanism is the motion of dislocations by sliding planes of the crystal [3]. This implies that interatomic interaction forces, directed transversely to the crystal sliding plane, can be overcome in case of the presence of local displacements number, determined by stresses periodic field in the lattice. This is strictly different from macroscopic shear process, during which all bonds are broken simultaneously (the Frenkel model). It seems obvious that with the help of dislocations total shear strain will be realized at the applying much lower external stress than for the process including simultaneous breakage of all atomic bonds by the sliding plane [3].

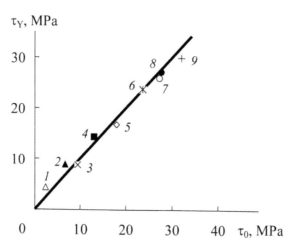

FIGURE 11.3   The relationship between theoretical τ_o and experimental τ_Y shear stresses at yielding for polytetrafluoroethylene (1), HDPE (2), polypropelene (3), polyamide-6 (4), poly(vinil chloride) (5), polyhydroxiester (6), PC (7), polysulfone (8) and PAr (9) [20].

Payerls and Nabarro [3] were the first who calculated the shear stress necessary for the dislocations motion, τ_{dm}. They used a sinusoidal approximation and deduced the expression for τ_{dm} as follows:

$$\tau_{dm} = \frac{2G}{1-v} e^{-2\pi a_i / b_i(1-v)}, \tag{5}$$

where v is the Poisson's ratio and parameters a_i and b_i are of the same meaning as in Fig. 11.2.

By substituting reasonable v value, for example, 0.35 [27], and assuming $a_i = b_i$, the following value for τ_{dm} is obtained: $\tau_{dm} = 2 \times 10^{-4}\ G$. Though for metals this value is higher than the observed τ_Y, it is much closer to them than the stress calculated using simple shear model (the Frenkel model, Fig. 11.1).

However, for polymers the situation is opposite: analogous calculation indicates that their τ_{dm} does not exceed 0.2 MPa, which is by two orders of magnitude, approximately, lower than the observed τ_Y values.

Let us consider further the free path length of dislocations, λ_d. As it is known for metals [3], in which the main role in plastic deformation belongs to the mobile dislocations, λ_d assesses as $\sim 10^4$ Å. For polymers, this parameter can be estimated as follows [28]:

$$\lambda_Y = \frac{\varepsilon_Y}{b\rho_d}, \tag{6}$$

where ε_Y is the yield strain, b is Burgers vector, ρ_d is the density if linear defects, determined according to the Eq. (1).

The value ε_Y assesses as ~ 0.10 [29] and the value of Burgers vector b can be estimated according to the equation [30]:

$$b = \left(\frac{60,7}{C_\infty} \right)^{1/2}, \text{Å}. \tag{7}$$

The values for different polymers, λ_d assessed by the Eq. (4.6) is about 2.5 Å. The same distance, which a segment passes at shearing, when it occupies the position, shown in Fig. 11.1b, that can be simply calculated from purely geometrical considerations. Hence, this assessment also indicates no reasons for assuming any sufficient free path length of dislocations in polymers rather than transition of a segment (or several segments) of macromolecule from one quasiequilibrium state to another [31].

It is commonly known [3, 25] that for crystalline materials Baily Hirsh relationship between shear stress, τ_Y, and dislocation density, ρ_d, is fulfilled:

$$\tau_Y = \tau_{in} + \alpha Gb\rho_d^{1/2},\tag{8}$$

where τ_{in} is the initial internal stress, α is the efficiency constant.

The Eq. (8) is also true for amorphous metals [21]. In Ref. [20] it was used for describing mechanical behavior of polymers on the example of these materials main classes representatives. For this purpose, the data for amorphous glassy PAr [32], semicrystalline HDPE [33] and cross-linked epoxy polymers of amine and anhydride curing types (EP) were used [34]. Different loading schemes were used: uniaxial tension of film samples [32], high-speed bending [33] and uniaxial compression [34]. In Fig 11.4, the relations between calculated and experimental values τ_Y for the indicated polymers are adduced, which correspond to the Eq. (4.8). As one can see, they are linear and pass through the coordinates origin (i.e., $\tau_{in} = 0$), but α values for linear and cross-linked polymers are different. Thus in the frameworks of the offered defect concept the Baily Hirsh relationship is also true for polymers. This means that dislocation analogs are true for any linear defect, distorting the material ideal structure and creating the elastic stresses field [20]. From this point of view high defectness degree of polymers will be noted: $\rho_d \approx 10^9 \div 10^{14}$ cm^{-2} for amorphous metals [21], and $\rho_d \approx 10^{14}$ cm^{-2} for polymers [28].

Hence, the stated above results indicate that in contrast with metals, for polymers realization of the Frenkel mechanism during yielding is much more probable rather than the defects motion (Fig. 11.1). This is due to the above-discussed (even diametrically opposed) differences in the structure of crystalline metals and polymers [1].

As it has been shown above using position spectroscopy methods [22], the yielding in polymers is realized in densely packed regions of their structure. Theoretical analysis within the frameworks of the plasticity fractal concept [35] demonstrates that the Poisson's ratio value in the yielding point, ν_Y, can be estimated as follows:

$$\nu_Y = \nu\chi + 0.5\,(1-\chi),\tag{9}$$

where ν is Poisson's ratio value in elastic strains field, χ is the relative fraction of elastically deformed polymer.

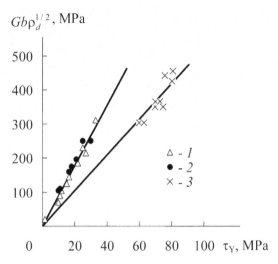

FIGURE 11.4 The relations between calculated and shear stress at yielding τ_Y, corresponding to the Eq. (4.8), for PAr (1), HDPE (2) and EP (3) [20].

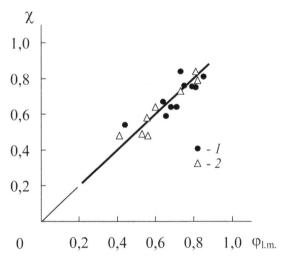

FIGURE 11.5 The relation between relative fractions of loosely packed matrix $\varphi_{l.m.}$ and probability of elastic state realization χ for PC (1) and PAr (2) [23].

In Fig. 11.5, the comparison of values χ and $\varphi_{l.m.}$ for PC and PAr is adduced, which has shown their good correspondence. The data of this figure assume that the loosely packed matrix can be identified as the elastic deformation region and clusters are identified as the region of inelastic (plastic) deformation [23]. These results prove the conclusion made in Ref. [22] about proceeding of inelastic deformation processes in dense packing regions of amorphous glassy polymers and indicate correctness of the plasticity fractal theory at their description.

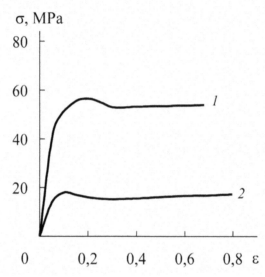

FIGURE 11.6   The stress-strain (σ–ε) diagrams for PAr at testing temperatures 293 (1) and 453 K (2) [24].

The yielding process of amorphous glassy polymers is often considered as their mechanical devitrification [36]. However, if typical stress–strain (σ–ε) plot for such polymers is considered (Fig. 11.6), then one can see, that behind the yield stress σ_Y the forced elasticity (cold flow) plateau begins and its stress σ_p is practically equal to σ_Y, that is, σ_p has the value of order of several tens MPa, whereas for devitrificated polymer this value is, at least, on the order of magnitude lower. Furthermore, σ_p is a function of the temperature of tests T, whereas for devitrificated polymer such dependence must be much weaker what is important, possess the opposite tendency (σ_p enhancement at T increase). This disparity is solved easily within the frameworks of the cluster model, where cold flow of polymers is associated with devitrivicated loosely packed matrix deformation, in which clusters "are floating." However, thermal devitrification of the loosely packed matrix occurs at the temperature T_g' – which is approximately 50 K lower than T_g. That is why it should be expected that amorphous polymer in the temperature range $T_g' \div T_g$ will be subjected to yielding under the application of even extremely low stress (of about 1 MPa). Nevertheless, as the plots in Fig. 11.6 show, this does not occur and σ–ε curve for PAr in $T_g' \div T_g$ range is qualitatively similar to the plot σ–ε at $T < T_g'$ (curve 1). Thus is should be assumed that devitrification of the loosely packed matrix is the consequence of the yielding process realization, but not its criterion. Taking into account realization of inelastic deformation process in the clusters (Fig. 11.5) one can suggest that the sufficient condition of yield in the polymer is the loss of stability by the local order regions in the external mechanical stress field, after which the deformation process proceeds without increasing the stress σ (at least, nominal one), contrary to deformation below the yield stress, where a monotonous increase of σ is observed (Fig. 11.6).

Now using the model suggested by the authors [37] one can demonstrate that the clusters lose their stability, when stress in the polymer reaches the macroscopic yield stress, σ_Y. Since the clusters are postulated as the set of densely packed collinear segments, and arbitrary orientation of cluster axes in relation to the applied tensile stress σ should be expected, then they can be simulated as "inclined plates" (IP) [37], for which the following expression is true [37]:

$$\tau_Y < \tau_{IP} = 24G_{cl}\varepsilon_0\left(1+v_2\right)/\left(2-v_2\right), \tag{10}$$

where τ_Y is the shear stress in the yielding point, τ_{IP} is the shear stress in IP (cluster), G_{cl} is the shear modulus, which is due to the clusters availability and determined from the plots.

Since the Eq. (10) characterizes inelastic deformation of clusters, the following can be accepted: $v_2 = 0.5$. Further on, under the assumption that $\tau_Y = \tau_{IP}$, the expression for the minimal (with regard to inequality in the left part of the Eq. (10)) proper strain ε_0^{min} is obtained [24]:

$$\varepsilon_0^{min} = \frac{\tau_Y}{\sqrt{24G_{cl}}}. \tag{11}$$

The condition for IP (clusters) stability looks as follows [37]:

$$q = \sqrt{\frac{3}{2}} \cdot \frac{\varepsilon_0}{\tau_Y} \left\{ \left|1 + \frac{\tilde{\varepsilon}_0}{\varepsilon_0}\right| - \sqrt{\frac{3}{8}} \frac{\tau_Y}{G_{cl}\varepsilon_0\left(1+v_2\right)} \right\}, \qquad (12)$$

where q is the parameter, characterizing plastic deformation, $\tilde{\varepsilon}_0$ is the proper strain of the loosely packed matrix.

The cluster stability violation condition is fulfillment of the following inequality [37]:

$$q \leq 0. \qquad (13)$$

Comparison of the Eqs. (4.12) and (4.13) gives the following criterion of stability loss for IP (clusters) [24]:

$$\left|1 + \frac{\tilde{\varepsilon}_0}{\varepsilon_0}\right| = \sqrt{\frac{3}{8}} \frac{\tau_Y^T}{G_{cl}\varepsilon_0\left(1+v_2\right)}, \qquad (14)$$

from which theoretical stress τ_Y (τ_Y^T) can be determined, after reaching of which the criterion (the Eq. (13)) is fulfilled.

To perform quantitative estimations, one should make two simplifying assumptions [24]. Firstly, for IP the following condition is fulfilled [37]:

$$0 \leq \sin^2\theta_{IP}\left(\tilde{\varepsilon}_0/\varepsilon_0\right) \leq 1, \qquad (15)$$

where θ_{IP} is the angle between the normal to IP and the main axis of proper strain.

Since for arbitrarily oriented IP (clusters) $\sin^2\theta_{IP} = 0.5$, then for fulfillment of the condition (the Eq. (15)) the assumption is enough. Secondly, the Eq. (11) gives the minimal value of ε_0, and for the sake of convenience of calculations parameters τ_Y and G_{cl} were replaced by σ_Y and E, respectively. E value is greater than the elasticity modulus E_{cl} due to the availability of clusters. That is why to compensate two mentioned effects the strain ε_0, estimated according to the Eq. (11), was twice increased. The final equation looks as follows [24]:

$$\varepsilon_0 \approx 0.64 \frac{\sigma_Y}{E} = 0.64\varepsilon_{cl}, \qquad (16)$$

where ε_{cl} is the elastic component of macroscopic yield strain [38], which corresponds to strains ε_0 and $\widetilde{\varepsilon}_0$ by the physical significance [37].

Combination of the Eqs. (14) and (16) together with the plots similar to the ones shown in Chapter 1, where from $E_{cl}(G_{cl})$ can be determined, allows to estimate theoretical yield stress σ_Y^T and compare it with experimental values σ_Y. Such comparison is adduced in Fig. 11.7, which demonstrates satisfactory conformity between σ_Y^T and σ_Y that proves the suggestion made in Ref. [24] and justifies the above-made assumptions.

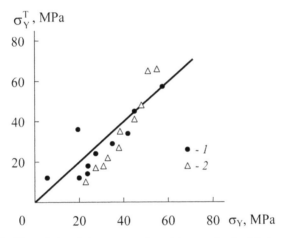

FIGURE 11.7   The relation between experimental σ_Y and calculated according to the equation (14) yield stress values for PAr (1) and PC (2) [24].

Hence, realization of the yielding process in amorphous glassy polymers requires clusters stability loss in the mechanical stress field, after which mechanical devitrification of the loosely packed matrix proceeds. Similar criterion was obtained for semicrystalline polymers [24].

As results obtained in Refs. [34, 39] have shown, the behavior of cross-linked polymers is just slightly different from the above-described one for linear PC and PAr. However, further progress in this field is quite difficult due to, at least, two reasons: excessive overestimation of the chemical cross-links role and the quantitative structural model absence. In Ref. [39], the yielding mechanism of cross-linked polymer has been offered, based on the application of the cluster model and the latest developments in the deformable solid body synergetics field [40] on the example of two already above-mentioned epoxy polymers of amine (EP-1) and anhydra-zide (EP-2) curing type.

Figure 11.8 shows the plots $\sigma-\varepsilon$ for EP-2 under uniaxial compression of the sample up to failure (curve 1) and at successive loading up to strain ε exceeding the yield strain ε_Y (curves 2–4). Comparison of these plots indicates consecutive lowering of the "yield tooth" under constant cold flow stress, σ_p. High values of σ_p assume corresponding values of stable clusters network density v_{cl}^{st}, which is much higher than the chemical cross-links network density v_c [34]. Thus though the behavior of a cross-linked polymer on the cold flow plateau is described within the frameworks of the rubber high-elasticity theory, the stable clusters network in this part of $\sigma-\varepsilon$ plots is preserved. The only process proceeding is the decay of instable clusters, determining the loosely packed matrix devitrification. This process begins at the stress equal to proportionality limit that correlates with the data from [41], where the action of this stress and temperature $T_2 = T_g$ is assumed analogous. The analogy between cold flow and glass transition processes is partial only: the only one component, the loosely packed matrix, is devitrivicated. Besides, complete decay of instable clusters occurs not in the point of yielding reaching at σ_Y, but at the beginning of cold flow plateau at σ_p. This can be observed from $\sigma-\varepsilon$ diagrams shown in Fig. 11.8. As a consequence, the yielding is regulated not by the loosely packed matrix devitrification, but by other mechanism. As it is shown above, as such mechanism the stability loss by clusters in the mechanical stress field can be assumed, which also follows from the well-known fact of derivative $d\sigma/d\varepsilon$ turning to zero in the yield point [42]. According to Ref. [40] critical shear strain γ_* leading to the loss of shear stability by a solid is equal to:

$$\gamma_* = \frac{1}{mn}, \tag{17}$$

where m and n are exponents in the Mie equation [27] setting the interconnection between the interaction energy and distance between particles. The value of parameter $1/mn$ can be expressed via the Poisson's ratio, v [27]:

$$\frac{1}{mn} = \frac{1-2v}{6(1+v)}. \tag{18}$$

From the Eq. (18) it follows [18]:

$$\frac{1}{mn} = \gamma_* = \frac{\sigma_Y}{E}.$$ (19)

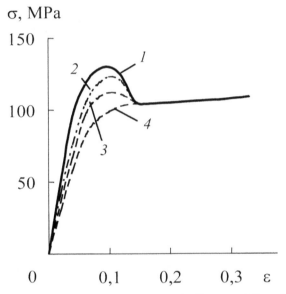

FIGURE 11.8 The stress–strain (σ–ε) diagrams at loading up to failure (1) and at cyclic load exertion (2–4) for EP-2: 2—the first loading cycle; 3—the second loading cycle; 4—the third loading cycle [39].

The equation (19) gives the strain value with no regard to viscoelastic effect, that is, diagram σ–ε deviation from linearity behind the proportionality limit. Taking into account that tensile strain is twice greater, approximately, than the corresponding shear strain [42], theoretical yield strain ε_Y^T, corresponding to the stability loss by a solid, can be calculated. In Fig. 11.9, the comparison of experimental ε_Y and ε_Y^T yield strain magnitudes is fulfilled. Approximate equality of these parameters is observed that assumes association of the yielding with the stability loss by polymers. More precisely, we are dealing with the stability loss by clusters, because parameter ν depends upon the cluster network density ν_{cl} and ε_Y value is proportional to ν_{cl} [32].

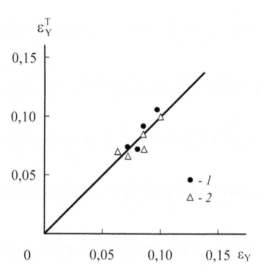

FIGURE 11.9   The relation between experimental ε_Y and theoretical ε_Y^T yield strain values for epoxy polymers EP-1 (1) and EP-2 (2) [39].

The authors [43] consider the possibility of intercommunication of polymers yield strain and these materials suprasegmental structure evolution, which is constituent part of hierarchical systems behavior [44, 45]. It is supposed that within the frameworks of this general concept polymer suprasegmental structures occupy their temporal and energetic "niches" in general hierarchy of real world structures [46]. As a structure quantitative model the authors [43] use a cluster model of polymers amorphous state structure [18, 24]. Ten groups of polymers, belonging to different deformation schemes in wide range of strain rates and temperatures, were used for obtaining possible greater results community. The yield strain ε_Y was chosen as the parameter, characterizing suprasegmental structures stability in the mechanical stresses field.

The Gibbs function of suprasegmental (cluster) structure self-assembly at temperature $T = T_g - \Delta T$ was calculated as follows [45]:

$$\Delta \widetilde{G}^m = \Delta S \Delta T , \tag{20}$$

where ΔS is entropy change in this process course, which can be estimated as follows [47]:

$$\Delta S = (3 \div 5) \times k \times f_g \times \ln f_g \tag{21}$$

In the Eq. (21) the coefficient $(3 \div 5)$ takes into account conformational molecular changes contribution to ΔS, k is Boltzmann constant, f_g is a polymer relative fluctuation free volume.

Let us consider now the results of the concept [44, 45] application to polymers yielding process description. The yielding can be considered as polymer structure loss of its stability in the mechanical stresses field and the yield strain ε_Y is measure of this process resistance. In Ref. [44] it is indicated that specific lifetime of suprasegmental structures t^{im} is connected with $\Delta \tilde{G}^m$ as follows:

$$t^{im} \sim \exp\left(-\Delta \tilde{G}^{im}/RT\right) \tag{22}$$

where R is the universal gas constant.

Assuming, that $t^{im} \sim t_Y$ (t_Y is time, necessary for yield stress σ_Y achievement) and taking into account, that $\varepsilon_Y \sim t_Y$, it can be written [43]:

$$\varepsilon_Y \sim \exp\left(-\Delta \tilde{G}^{im}/RT\right). \tag{23}$$

Let us note, that the Eq. (23) correctness in reference to polymers means, that yielding process in them is controlled by supra segmental structures thermodynamically stability.

In Fig. 11.10, the dependence of ε_Y on $\exp\left(-\Delta \tilde{G}^{im}/RT\right)$, corresponding to the Eq. (23), for all groups of the considered in Ref. [43] polymers. Despite definite (and expected) data scattering it is obvious, that all data break down into two branches, the one of which is approximated by a straight line. Such division reasons are quite obvious: the data with negative value of $\Delta \tilde{G}^m$ cover one (right) branch and with positive ones – the other (left) one. The last group consists of semicrystalline polymers with devitrificated at testing temperature loosely packed matrix (polytetrafluoroethylene, polyethylenes, polypropylene). The cluster model [18, 24] postulates thermo-fluctuation character of clusters formation and their decay at $T \geq T_g$. Therefore such clusters availability in the indicated semicrystalline polymers devitrificated amorphous phase has quite another origin, namely, it is due to amorphous chains tightness in crystallization process [48]. In practice this effect results to the condition $\Delta\tilde{G}^m > 0$ realization, which in the low-molecular substances case belongs to hypothetical non-existent in reality transitions "an overheated liquid ® solid body" [45].

Let us note one more feature, confirming existence reality of Fig. 11.10 plot left branch. At present ε_Y increases at T growth for polyethylenenes and ε_Y decrease at the same conditions—for amorphous glassy polymers are well known [18]. One can see easily, that this experimental fact is explained completely by two branches availability in the plot of Fig. 11.10, which confirms again existence reality of suprasegmental structures, which are in quasiequilibrium with "free" segments. This quasiequilibrium is characterized by $\Delta\tilde{G}^{im}$ so, that in this case $\Delta\tilde{G}^{im} > 0$.

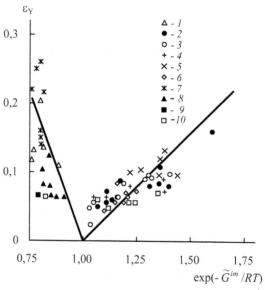

FIGURE 11.10 The correlation of yield strain ε_Y and parameter $\exp\left(-\Delta\tilde{G}^{im}/RT\right)$ for polymers with devitrificated (1, 7, 8, 9) and glassy (2–6, 10) loosely packed matrix [43].

As it was to be expected the value $\varepsilon_Y = 0$ at $\exp\left(-\Delta\tilde{G}^{im}/RT\right)$ or $\Delta\tilde{G}^{im} = 0$. The last condition is achieved at $T = T_g$, where the yield strain is always equal to zero. Let us also note, that the data for semicrystalline polymers with vitrificated amorphous phase (polyamide-6, poly(ethylene terephthalate)) cover the right branch of the Fig. 11.10 plot. This means, that suprasegmental structures existence ($\Delta \tilde{G}^{im} > 0$) is due just to devitrificated amorphous phase availability, but no crystallinity. At $T < T_g$ amorphous phase viscosity, increases sharply and amorphous chains tightness cannot be exercised its action, displacing macromolecules parts, owing to that local order formation has thermofluctuation character. The attention is paid to the obtained plot community, if to remember, that the values t_Y for impact and quasistatic tests are differed by five orders. This circumstance is explained simply enough, since the value ε_Y can be written as follows [43]:

$$\varepsilon_Y = t_Y \dot{\varepsilon}, \tag{24}$$

where $\dot{\varepsilon}$ is strain rate and substitution of the Eq. (24) in the Eq. (23) shows, that in the right part of the latter the factor $\dot{\varepsilon}^{-1}$ appears, which for the considered loading schemes changes by about five orders.

The Gibbs specific function notion for nonequilibrium phase transition "overcooled liquid ® solid body" is connected closely to local order notion (and, hence, fractality notion, see Chapter 1), since within the frameworks of the cluster model the indicated transition is equivalent to cluster formation start. The dependence of clusters relative fraction φ_{cl} on the value $\left| \Delta \tilde{G}^{im} \right|$ for PC and PAr is adduced. As one can see, this dependence is linear, φ_{cl} growth at $\left| \Delta \tilde{G}^{im} \right|$ increasing is observed and at $\left| \Delta \tilde{G}^{im} \right| = 0$ (i.e., for the selected standard temperature $T = T_g$) the cluster structure complete decay ($\varphi_{cl} = 0$) occurs.

The adduced above results can give one more, at least partial, explanation of "cell's effect." As it has been shown in Ref. [18], the following approximate relationship exists between ε_Y and Grüneisen parameter γ_L:

$$\varepsilon_Y = \frac{1}{2\gamma_L}. \tag{25}$$

Using this relationship and the plots of Fig. 11.10, it is easy to show, that the decrease and, hence, φ_{cl} reduction results to γ_L growth, characterizing intermolecular bonds anharmonicity level. This parameter shows, how fast intermolecular interaction weakens at external (e.g., mechanical one [49]) force on polymer and the higher γ_L the faster intermolecular interaction weakening occurs at other equal conditions. In other words, the greater $\left|\Delta\tilde{G}^{im}\right|$ and φ_{cl}, the smaller γ_L and the higher polymer resistance to external influence. In Fig. 11.11, the dependence of γ_L on mean number of statistical segments per one cluster n_{cl}, which demonstrates clearly the said above.

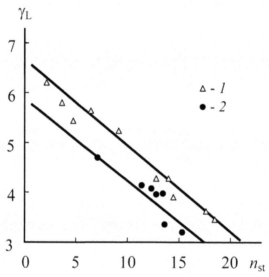

FIGURE 11.11 The dependence of Grüneisen parameter γ_L on mean number of statistical segments per one cluster n_{cl} for PC (1) and PAr (2) [43].

Hence, the stated above results shown that polymer yielding process can be described within the frameworks of the macrothermodynamical model. This is confirmed by the made in Ref. [44] conclusion about thermodynamically factor significance in those cases, when quasiequilibrium achievement is reached by mechanical stresses action. The existence possibility of structures with $\left|\Delta\tilde{G}^{im}\right| > 0$ (connected with transition "overheated liquid ® solid body" [45] was shown and, at last, one more possible treatment of "cell's effect" was given within the frameworks of intermolecular bonds anharmonicity theory for polymers [49].

If to consider the yielding process as polymer mechanical devitrification [36], then the same increment of fluctuation free volume f_g is required for the strain ε_Y achievement. This increment Δf_g can be connected with ε_Y as follows [50]:

$$\Delta f_g = \varepsilon_Y(1-2v). \tag{26}$$

Therefore, f_g decreases results to Δf_g growth and respectively, ε_Y enhancement.

Let us consider, which processes result to necessary for yielding realization fluctuation free volume increasing. Theoretically (within the frameworks of polymers plasticity fractal concept [35] and experimentally (by positrons annihilation method [22]) it has been shown, that the yielding process is realized in densely packed regions of polymer. It is obvious, that the clusters will be such regions in amorphous glassy polymer and in semicrystalline one—the clusters and crystallites [18]. The Eq. (9) allows to estimate relative fraction χ of polymer, which remains in the elastic state.

In Fig. 11.12, the temperature dependence of χ for HDPE is shown, from which one can see its decrease at T growth. The absolute values χ change within the limits of 0,516 ÷ 0,364 and the determined by polymer density crystallinity degree K for the considered HDPE is equal to 0,687 [51]. In other words, the value χ in all cases exceeds amorphous regions fraction and this means the necessity of some part crystallites melting for yielding process realization. Thus, the conclusion should be made, that a semicrystalline HDPE yielding process includes its crystallites partial mechanical melting (disordering). For the first time Kargin and Sogolova [52] made such conclusion and it remains up to now prevalent in polymers mechanics [53]. The concept [35] allows to obtain quantitative estimation of crystallites fraction χ_{cr}, subjecting to partial melting − recrystallization process, subtracting amorphous phase fraction in HDPE from χ. The temperature dependence of χ_{cr} was also shown in Fig. 11.12, from which one can see, that χ_{cr} value is changed within the limits of 0.203   0.051, decreasing at T growth.

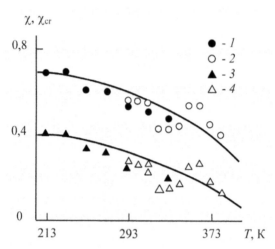

FIGURE 11.12 The dependences of elastically deformed regions fraction χ (1, 2) and crystallites fraction, subjected to partial melting, χ_{cr} (3, 4) on testing temperature T in impact (1, 3) and quasistatic (2, 4) tests for HDPE [51].

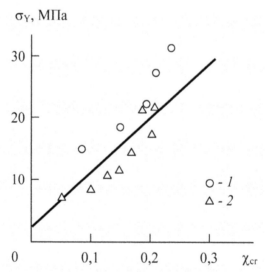

FIGURE 11.13 The dependence of yield stress σ_Y on crystallites fraction, subjecting to partial melting χ_{cr}, in impact (1) and quasistatic (2) tests for HDPE [51].

It is natural to assume, that the yield stress is connected with parameter χ_{cr} as follows: the greater χ_{cr}, the larger the energy, consumed for melting and the higher σ_Y. The data of Fig. 11.13 confirm this assumption and the dependence $\sigma_Y(\chi_{cr})$ is extrapolated to finite σ_Y value, since not only crystallites, but also clusters participate in yielding process. As it was to be expected [54], the crystalline regions role in yielding process realization is much larger, than the amorphous ones.

Within the frameworks of the cluster model [55] it has been assumed, that the segment joined to cluster means fluctuation free volume microvoid "shrinkage" and vice versa. In this case the microvoids number ΔN_h, forming in polymer dilation process, should be approximately equal to segments number ΔN_f, subjected to partial melting process. These parameters can be estimated by following methods. The value ΔN_h is equal to [51]:

$$\Delta N_h = \frac{\Delta f_g}{V_h}, \tag{27}$$

where V_h is free volume microvoid volume, which can be estimated according to the kinetic theory of free volume [27].

Macromolecules total length L per polymer volume unit is estimated as follows [56]:

$$L = \frac{1}{S}, \tag{28}$$

where S is macromolecule cross-sectional area.

The length of macromolecules L_{cr}, subjected to partial melting process, per polymer volume unit is equal to [51]:

$$L_{cr} = L\chi_{cr} \tag{29}$$

Further the parameter ΔN_χ can be calculated [51]:

$$\Delta N_\chi = \frac{L_{cr}}{l_{st}} = \frac{L\chi_{cr}}{l_0 C_\infty}. \tag{30}$$

The comparison of ΔN_h and ΔN_Y values is adduced in Fig. 11.14, from which their satisfactory correspondence follows. This is confirmed by the conclusion, that crystallites partial mechanical melting (disordering) is necessary for fluctuation free volume f_g growth up to the value, required for polymers mechanical devitrification realization [51].

Whenever work is done on a solid, there is also a flow of heat necessitated by the deformation. The first law of thermodynamics:

$$dU = dQ + dW \qquad (31)$$

states that the internal energy change dU of a sample is equal to the sum of the work dW performed on the sample and the heat flow dQ into the sample. This relation is valid for any deformation, whether reversible or irreversible. There are two thermodynamically irreversible cases for which dQ and dW are equal by absolute value and opposite by sign: uniaxial deformation of a Newtonian fluid and ideal elastic-plastic deformation. For amorphous glassy polymers deformation has essentially different character: the ratio $dQ/dW \neq 1$ and changes within the limits of 0.36 0.75 depending on testing condition [57]. In other words, the thermodynamically ideal plasticity is not realized for these materials. The reason of this effect is thermodynamically nonequilibrium of polymers structure. Within the frameworks of the fractal analysis it has been shown that it results to polymers yielding process realization not only in entire sample volume, but also in its part (see the Eq. (9)) [35]. Besides, it has been demonstrated experimentally and theoretically that amorphous glassy polymer structural component, in which the yielding is realized, is densely packed local order regions (clusters) [22, 23].

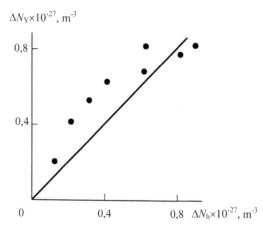

FIGURE 11.14 The relation between increment of segments number ΔN_Y in crystallites, subjecting to partial melting, and increment of free volume microvoids number ΔN_h, which is necessary for yielding process realization, for HDPE [51].

Lately the mathematical apparatus of fractional integration and differentiation [58, 59] was used for fractal objects description, which is amorphous glassy polymers structure. It has been shown [60] that Kantor's set fractal dimension coincides with an integral fractional exponent, which indicates system states fraction, remaining during its entire evolution (in our case deformation). As it is known [61], Kantor's set ("dust") is considered in one-dimensional Euclidean space ($d = 1$) and therefore its fractal dimension obey the condition $d_f \leq 1$. This means, that for fractals, which are considered in Euclidean spaces with $d > 2$ ($d = 2, 3, \ldots$) the fractional part of fractal dimension should be taken as fractional exponent ν_{fr} [62, 63]:

$$\nu_{fr} = d_f - (d - 1). \tag{32}$$

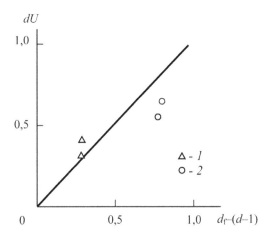

FIGURE 11.15 The dependence of relative fraction of latent energy dU on fractional exponent $\nu_{fr} = d_f - (d - 1)$ value for PC (1) and PMMA (2) [64].

The value ν_{fr} characterizes that states (structure) part of system (polymer), which remains during its entire evolution (deformation). In Fig. 11.15, the dependence of latent energy fraction dU at PC and poly(methyl methacrylate) (PMMA) deformation on $\nu_{fr} = d_f - (d - 1)$ [64] is shown. The value dU was estimated as $(W-Q)/W$. In Fig. 11.15, the theoretical dependence $dU(\nu_{fr})$ is adducted, plotted according to the conditions $dU = 0$ at $\nu_{fr} = 0$ or $d_f = 2$ and $dU = 1$ at $\nu_{fr} = 1$ or $d_f = 3$ ($d = 3$), that is, at ν_{fr}

or d_f limiting values [64]. The experimental data correspond well to this theoretical dependence, from which it follows [64]:

$$dU = v_{fr} = d_f - (d - 1). \tag{33}$$

Let us consider two limiting cases of the adduced in Fig. 11.15 dependence at $v_{fr} = 0$ and 1.0, both at $d = 3$. In the first case ($d_f = 2$) the value $dU = 0$ or, as it follows from dU definition (the equation (31)), $dW = dQ$ and polymer possesses an ideal elastic-plastic deformation. Within the frameworks of the fractal analysis $d_f = 2$ means, that $\varphi_{cl} = 1.0$, that is, amorphous glassy polymer structure represents itself one gigantic cluster. However, as it has been shown above, the condition $d_f = 2$ achievement for polymers is impossible in virtue of entropic tightness of chains, joining clusters, and therefore $d_f > 2$ for real amorphous glassy polymers. This explains the experimental observation for the indicated polymers: $dU \neq 0$ or $|dW| \neq |dQ|$ [57]. At $v_{fr} = 1,0$ or $d_f = d = 3$ polymers structure loses its fractal properties and becomes Euclidean object (true rubber). In this case from the plot of Fig. 11.15 $dU = 1.0$ follows. However, it has been shown experimentally, that for true rubbers, which are deformed by thermodynamically reversible deformation $dU = 0$. This apparent discrepancy is explained as follows [64]. Fig. 11.15 was plotted for the conditions of inelastic deformation, whereas at $d_f = d$ only elastic deformation is possible. Hence, at $v_{fr} = 1.0$ or $d_f = d = 3$ deformation type discrete (jump-like) change occurs from $dU ® 1$ up to $dU = 0$. This point becomes an initial one for fractal object deformation in Euclidean space with the next according to the custom dimension $d = 4$, where $3 \leq d_f \leq 4$ and all said above can be repeated in reference to this space: at $d = 4$ and $d_f = 3$ the value $v_{fr} = 0$ and $dU = 0$. Let us note in conclusion that exactly the exponent v_{fr} controls the value of deformation (fracture) energy of fractal objects as a function of process length scale. Let us note that the equality $dU = \varphi_{1.m.}$ was shown, from which structural sense of fractional exponent in polymers inelastic deformation process follows: $v_{fr} = \varphi_{1.m.}$ [64].

Mittag-Lefelvre function [59] usage is one more method of a diagrams $\sigma - \varepsilon$ description within the frameworks of the fractional derivatives mathematical calculus. A nonlinear dependences, similar to a diagrams $\sigma - \varepsilon$ for polymers, are described with the aid of the following equation [65]:

$$\sigma(\varepsilon) = \sigma_0 \left[1 - E_{v_{fr},1} \left(-\varepsilon^{v_{fr}} \right) \right],\tag{34}$$

where σ_0 is the greatest stress for polymer in case of linear dependence $\sigma(\varepsilon)$ (of ideal plasticity), is the Mittag-Lefelvre function [65]:

$$E_{v_{fr},1} \left(-\varepsilon^{v_{fr}} \right) = \sum_{k=0}^{\infty} \frac{\varepsilon^{v_{fr}k}}{\tilde{A} \left(v_{fr}k + \beta \right)},\ \text{nfr} > 0,\ b > 0,\tag{35}$$

where Γ is Eiler gamma-function.

As it follows from the Eq. (34), in the considered case $\beta = 1$ and gamma-function is calculated as follows [40]:

$$\tilde{A} \left(v_{fr}k + 1 \right) = \sqrt{\frac{\pi}{2}} \left(v_{fr}k - \varepsilon^{v_{fr}k} \right)^{v_{fr}k - \varepsilon^{-v_{fr}}} e^{-\left(v_{fr}k - \varepsilon^{v_{fr}} \right)}.\tag{36}$$

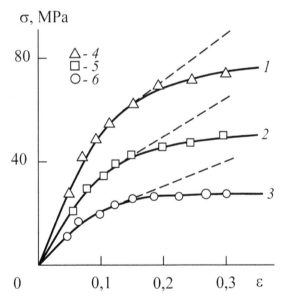

FIGURE 11.16 The experimental (1 ÷ 3) and calculated according to the equations (4.34) ÷ (4.36) (4 ÷ 6) diagrams σ–ε for PAr at $T = 293$ (1, 4), 353 (2, 5) and 433 K (3, 6). The shaded lines indicate calculated diagrams σ–ε for forced high-elasticity part without v_{fr} change [66].

In Fig. 11.16, the comparison of experimental and calculated according to the Eqs. (34) ÷ (36) diagrams $\sigma - \varepsilon$ for PAr at three testing temperatures is adduced. The values σ_o was determined as the product $E\varepsilon_Y$ [66]. As it follows from the data of Fig. 11.16, the diagrams $\sigma - \varepsilon$ on the part from proportionality limit up to yield stress are well described well within the frameworks of the Mittag-Lefelvre function. Let us note that two neces-sary for these parameters (σ_o and v_{fr}) are the function of polymers struc-tural state, but not filled parameters. This is a principal question, since the usage in this case of empirical fitted constants, as, for example, in Ref. [67], reduces significantly using method value [60, 65].

In the initial linear part (elastic deformation) calculation according to the Eq. (34) was not fulfilled, since in it deformation is submitted to Hooke law and, hence, is not nonlinear. At stresses greater than yield stress (high-elasticity part) calculation according to the Eq. (34) gives stronger stress growth (stronger strain hardening), than experimentally observed (that it has been shown by shaded lines in Fig. 11.16). The experimental and theo-retical dependences $\sigma - \varepsilon$ matching on cold flow part within the frame-works of the equation (34) can be obtained at supposition $v_{fr} = 0.88$ for $T = 293$ K and $v_{fr} = 1.0$ for the two remaining testing temperatures. This ef-fect explanation was given within the frameworks of the cluster model of polymers amorphous state structure [39], where it has been shown that in a yielding point small (instable) clusters, restraining loosely packed matrix in glassy state, break down. As a result of such mechanical devitrifica-tion glassy polymers behavior on the forced high-elasticity (cold flow) plateau is submitted to rubber high-elasticity laws and, hence, $d_f \circledR d = 3$ [68]. The stress decay behind yield stress, so-called "yield tooth", can be described similarly [66]. An instable clusters decay in yielding point re-sults to clusters relative fraction φ_{cl} reduction, corresponding to d_f growth and v_{fr} enhancement (the Eq. (33)) and, as it follows from the Eq. (34), to stress reduction. Let us note in conclusion that the offered in Ref. [69] techniques allow to predict parameters, which are necessary for diagrams $\sigma-\varepsilon$ description within the frameworks of the considered method, that is, E, ε_Y and d_f.

In Ref. [70], it has been shown that rigid-chain polymers can be have a several substates within the limits of glassy state. For polypiromellithi-mide three such substates are observed on the dependences of modulus

$d\sigma/d\varepsilon$, determined according to the slope of tangents to diagram $\sigma-\varepsilon$, on strain ε [70]. However, such dependences of $d\sigma/d\varepsilon$ on ε have much more general character: in Fig. 11.17 three similar dependences for PC are adduced, which were plotted according to the data in Ref. [49]. If in Ref. [70], the transition from part I to part II corresponds to polypiromellithimide structure phase change with axial crystalline structure formation [71], then for the adduced in Fig. 11.17 similar dependences for PC this transition corresponds to deformation type change from elastic to inelastic one (proportionality limit [72]). The dependences $(d\sigma/d\varepsilon)(\varepsilon)$ community deserves more intent consideration that was fulfilled by the authors [73] within the frameworks of fractal analysis.

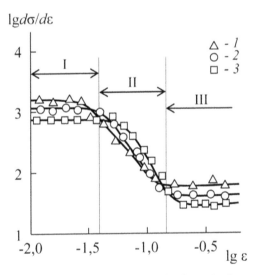

FIGURE 11.17   The dependences of modulus $d\sigma/d\varepsilon$, determined according to the slope of tangents to diagram $\sigma-\varepsilon$, on strain ε in double logarithmic coordinates for PC at $T = 293$ (1), 343 (2) and 373 K (3) [73].

The dependence of physical fractal density ρ on measurement scale L in double logarithmic coordinates was shown. For $L < L_{min}$ and $L > L_{max}$ Euclidean behavior is observed and within the range of $L = L_{min} \div L_{max}$ – fractal one [40]. Let us pay attention to the complete analogy of the plots of Fig. 11.17.

There is one more theoretical model allowing nonstandard treatment of the shown in Fig. 11.17 dependences $d\sigma/d\varepsilon(\varepsilon)$. Kopelman was offered the fractal descriptions of chemical reactions kinetics, using the following simple relationship [74]:

$$k \sim t^{-h}, \qquad\qquad (37)$$

where k is reaction rate, t is its duration, h is reactionary medium nonhomogeneity (heterogeneity) exponent ($0 < h < 1$), which turns into zero only for Euclidean (homogeneous) mediums and the relationship (37) becomes classical one: $k = $ const.

From the Eq. (37) it follows, that in case $h \neq 0$, that is, for heterogeneous (fractal) mediums the reaction rate k reduces at reaction proceeding. One should attention to qualitative analogy of curves σ–ε and the dependences of conversion degree on reaction duration $Q(t)$ for a large number of polymers synthesis reactions [75]. Still greater interest for subsequent theoretical developments presents complete qualitative analogy of diagrams σ–ε and strange attractor trajectories, which can be have "yield tooth", strain hardening and so on [76].

If to consider deformation process as polymer structure reaction with supplied, from outside mechanical energy to consider, then the modulus $d\sigma/d\varepsilon$ will be k analog (Fig 11.17). The said above allows to assume, that deformation on parts I and III (elasticity and cold flow) proceeds in Euclidean space and on part II (yielding) – in fractal one.

The fractal analysis main rules usage for polymers structure and properties description [68, 77] allows to make quantitative estimation of measurement scale L change at polymer deformation. There are a several methods of such estimation and the authors [73] use the simplest from them as ensuring the greatest clearness. As it was noted in Chapter 1, the self-similarity (fractality) range of amorphous glassy polymers structure coincides with cluster structure existence range: the lower scale of self-similarity corresponds to statistical segment length l_{st} and the upper one-to distance between clusters R_{cl}. The simplest method of measurement scale L estimation is the usage of well-known Richardson equation.

In the case of affine deformation the value R_{cl} will be changed proportionally to drawing ratio λ [70]. This change value can be estimated from the equation [79]:

$$R_{cl}\lambda = l_{st}\frac{2(1-v)}{(1-2v)}.$$
(38)

From the Eq. (38) v increase follows—the d_f increase at drawing ratio λ growth. In its turn, the value λ is connected with the strain ε by a simple relationship (in the case of affine deformation) [80]:

$$\lambda = 1 + \varepsilon.$$
(39)

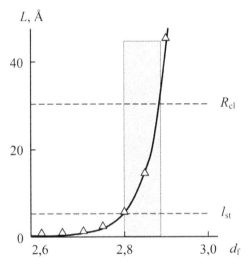

FIGURE 11.18 The dependence of measurement scale L on structure fractal dimension d_f for PC at $T = 293$ K. Horizontal shaded lines indicate nondeformed PC structure self-similarity boundaries (l_{st} and R_{cl}) and the shaded region—deformation fractal behavior range [73].

In Fig. 11.18 the dependence of L on d_f is adduced, which is calculated according to the Eqs. (38) and (39) combination and at the condition, that the parameters $C_\infty d_f$ are connected with each other as follows [18]:

$$C_\infty = \frac{2d_f}{d(d-1)(d-d_f)} + \frac{4}{3}.$$
(40)

As it follows from the data of Fig. 11.18, L growth at d_f increase is observed and within the range of $d_f \approx 2.80 \div 2.89$ polymer deformation proceeds in fractal space. At $d_f > 2.90$ the deformation space transition from fractal to Euclidean one is observed (PC yielding is achieved at $d_f = 2.85$ [23]) and structure PC approaching to true rubber state ($d_f = d = 3$) induces very fast L growth. Let us consider the conditions of transition from part II to part III (Fig. 11.17). The value R_{cl} by $R_{cl}\lambda$ according to the indicated above reasons it assumption $D_{ch} = 1.0$, that is, part of chain, stretched completely between clusters, let us obtain [73]:

$$\frac{L_{cl}}{l_{st}} = \frac{R_{cl}\lambda}{l_{st}}, \tag{41}$$

That is, drawing ratio critical value λ_{cr}, corresponding to the transition from part II to part III (from fractal behavior to Euclidean one) or the transition from yielding to cold flow, is equal to [73]:

$$\lambda_{cr} = \frac{L_{cl}}{R_{cl}l_{st}}, \tag{42}$$

that corresponds to the greatest attainable molecular draw [81].

For PC at $T = 293$ K and the indicated above values L_{cl} and R_{cl} λ_{cr} will be equal to 2.54. Taking into account, that drawing ratio at uniaxial deformation $\lambda"_{cr} = \lambda_{cr}^{1/3}$, let us obtain $\lambda"_{cr} = 1.364$ or critical value of strain of transition. Let us note that within the frameworks of the cluster model of polymers amorphous state structure [18] chains deformation in loosely packed matrix only is assumed and since the Eq. (42) gives molecular drawing ratio, which is determined in the experiment according to the relationship [81]:

$$\varepsilon_{cr} = \varepsilon"_{cr}(1 - \phi_{cl}). \tag{43}$$

Let us obtain $\varepsilon_{cr} = 0.117$ according to the eq. (4.43). The experimental value of yield strain ε_y for PC at $T = 293$ K is equal to 0.106. This means, that the transition to PC cold flow begins immediately beyond yield stress, which is observed experimentally [23].

Therefore, the stated above results show, that the assumed earlier sub-states within the limits of glassy state are due to transitions from deformation in Euclidean space to deformation in fractal space and vice versa. These transitions are controlled by deformation scale change, induced by external load (mechanical energy) application. From the physical point of view this postulate has very simple explanation: if size of structural element, deforming deformation proceeding, hits in the range of sizes L_{min}–L_{max} (Fig. 11.2), then deformation proceeds in fractal space, if it does not hit – in Euclidean one. In part I intermolecular bonds are deformed elastically on scales of $3 \div 4$Å ($L < L_{min}$), in part II—cluster structure elements with sizes of order of $6 \div 30$Å [18, 24] ($L_{min} < L < L_{max}$) and in part III chains fragments with length of L_{cl} or of order of several tens of Ånströms ($L > L_{max}$). In Euclidean space the dependence $\sigma - \varepsilon$ will be linear ($d\sigma/d\varepsilon$ = const) and in fractal one – curvilinear, since fractal space requires deformation deceleration with time. The yielding process realization is possible only in fractal space. The stated model of deformation mechanisms is correct only in the case of polymers structure presentation as physical fractal.

In the general case, polymers structure is multifractal for behavior description of which in deformation process in principle its three dimensions knowledge is enough: fractal (Hausdorff) dimension d_f, informational one d_1 and correlation one d_c [82]. Each from the indicated dimension describes multifractal definite properties change and these dimensions combined application allows to obtain more or less complete picture of yielding process [73].

As it is known [83], a glassy polymers behavior on cold flow plateau (part III in Fig. 11.17) is well described within the frameworks of the rubber high-elasticity theory. In Ref. [39], it has been shown that this is due to mechanical devitrification of an amorphous polymers loosely packed matrix. Besides, it has been shown [82, 84] that behavior of polymers in rubber-like state is described correctly under assumption, that their structure is a regular fractal, for which the identity is valid:

$$d_1 = d_c = d_f. \tag{44}$$

A glassy polymers structure in the general case is multifractal [85], for which the inequality is true [82, 84]:

$$d_c < d_1 < d_f. \tag{45}$$

Proceeding from the said above and also with appreciation of the known fact, that rubbers do not have to some extent clearly expressed yielding point the authors [73] proposed hypothesis, that glassy polymer structural state changed from multifractal up to regular fractal, that is, criterion (44) fulfillment, was the condition of its yielding state achievement. In other words, yielding in polymers is realized only in the case, if their structure is multifractal, that is, if it submits to the inequality (45).

Let us consider now this hypothesis experimental confirmations and dimensions d_1 and d_c estimation methods in reference to amorphous glassy polymers multifractal structure. As it is known [82], the informational dimension d_1 characterizes behavior Shennone informational entropy $I(\varepsilon)$:

$$I(\varepsilon) = \sum_{i}^{M} P_i \ln P_i,$$ (46)

where M is the minimum number of d-dimensional cubes with side ε, necessary for all elements of structure coverage, P_i is the event probability that structure point belongs to i-th element of coverage with volume ε^d.

In its turn, polymer structure entropy change ΔS, which is due to fluctuation free volume f_g, can be determined according to the Eq. (21). Comparison of the Eqs. (21) and (46) shows that entropy change in the first from them is due to f_g change probability and to an approximation of constant the values $I(\varepsilon)$ and ΔS correspond to each other. Further polymer behavior at deformation can be described by the following relationship [84]:

$$\Delta S = -c \left(\sum_{j=1}^{d} \lambda_i^{d_1} - 1 \right),$$ (47)

where c is constant, λ_i is drawing ratio.

Hence, the comparison of the Eqs. (21), (46) and (47) shows that polymer behavior at deformation is defined by change f_g exactly, if this parameter is considered as probabilistic measure. Let us remind, that f_c such definition exists actually within the frameworks of lattice models, where this parameter is connected with the ratio of free volume microvoids number N_h and lattice nodes number N (N_h/N) [49]. The similar definition is given and for P_i in the Eq. (46) [86].

The value d_1 can be determined according to the following equation [82]:

$$d_1 = \lim_{\varepsilon \to 0} \left[\frac{\sum_{i=1}^{M} P_i(\varepsilon) \ln P_i(\varepsilon)}{\ln \varepsilon} \right]. \tag{48}$$

Since polymer structure is a physical fractal (multifractal), then for it fractal behavior is observed only in a some finite range of scales (please give the reference on the Chapter 1) and the statistical segment length l_{st} is accepted as lower scale. Hence, assuming $P_i(\varepsilon) = f_g$ and $\varepsilon ® l_{st}$ the Eq. (48) can be transformed into the following one [73]:

$$d_1 = c_1 \frac{Rf_g \ln f_g}{\ln l_{st}}, \tag{49}$$

where c_1 is constant.

As it is known [23], in yielding point for PC $d_f = 2.82$ (or $v = 0.41$), that allows to determine the value f_g according to the previous equations and constant c_1, from the condition (the Eq. (44)). Then, using the known equality $\lambda_1 = 1 + \varepsilon_1$, similar to the Eq. (39), the yield strain ε_Y theoretical value can be calculated, determining constant c by matching method and the value Δf_g for ΔS calculation, as values f_g difference for nondeformed polymer and polymer in yielding point. Since in Ref. [73] thin films are subjected to deformation, then as the first approximation $d = 1$ is assumed in the Eq. (47) sum sign. In Fig. 11.19, comparison of the temperature dependences of experimental and calculated by the indicated method values ε_Y for PC is adduced. As one can see, the good correspondence of theory and experiment is obtained by both the dependences $\varepsilon_Y(T)$ course and ε_Y absolute values and discrepancy of experimental and theoretical values ε_Y at large T is due to the fact, that for calculation simplicity the authors [73] did not take into consideration transverse strains influence on value ε_Y in the Eq. (47).

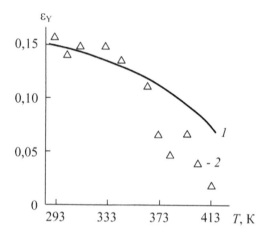

FIGURE 11.19 The comparison of experimental (1) and calculated according to the equation (47) temperature dependences of yield strain ε_Y for PC [73].

Williford [87] proposed, that the value d_I of multifractal corresponded to its surface dimension (the first subfractal) – either sample surface or fracture surface. For this supposition checking the authors [73] calculate a fracture surface fractal dimension for brittle (d_{fr}^{br}) and ductile (d_{fr}^{duc}) failure types according to the equations [40]:

$$d_{fr}^{br} = \frac{10(1+v)}{7-3v} \tag{50}$$

and

$$d_{fr}^{duc} = \frac{2(1+4v)}{1+2v}. \tag{51}$$

In Fig. 11.20, comparison of the temperature dependences of d_I, d_{fr}^{br} and d_{fr}^{duc} for PC is adduced. As one can see, at low T ($T < 373$ K) the value d_I corresponds to d_{fr}^{br} well enough and at higher temperatures ($T > 383$ K) it is close to d_{fr}^{duc}. In Fig. 11.20, the shaded region shows the temperature range corresponding to the brittle-ductile transition for PC [73]. It is significant that this interval beginning ($T = 373$ K) coincides with loosely packed matrix devitrification temperature T_g', which is approximately 50

K lower than polymer glass transition temperature T_g [18], for PC equal to ~ 423 K [88].

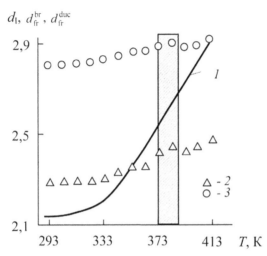

FIGURE 11.20 Comparison of the temperature dependences of informational dimension d_1 (1), fractal dimensions of fracture surface at brittle d_{fr}^{br} (2) and ductile d_{fr}^{duc} (3) failure for PC. The temperature range of brittle-ductile transition is shown by shaded region [73]

The correlation dimension d_c is connected with multifractal structure internal energy U [61] and it can be estimated according to the equation [82]:

$$\Delta U = -c_2 \left(\lambda_F^{d-d_c} \right),$$ (52)

where ΔU is internal energy change in deformation process, c_2 is constant, λ_F is macroscopic drawing ratio, which in the case of uniaxial deformation is equal to λ_1. The value λ_F is determined as follows [82]:

$$\lambda_F = \lambda_1 \lambda_2 \lambda_3,$$ (53)

where λ_2 and λ_3 are transverse drawing rations, connected with λ_1 by the simple relationships [82]:

$$\lambda_2 = 1 + \varepsilon_2,$$ (54)

$$\lambda_3 = 1 + \varepsilon_3, \tag{55}$$

$$\varepsilon_2 = \varepsilon_3 = v\varepsilon_1. \tag{56}$$

The temperature dependence of d_c can be calculated, as and earlier, assuming from the condition (the Eq. (44)) that at yielding $d_c = d_f = 2.82$, estimating ΔU as one half of product $\sigma_Y \varepsilon_Y$ (with appreciation of practically triangular form of curve σ—ε up to yield stress) and determining the constant c_2 by the indicated above mode. Comparison of the temperature dependences of multifractal three characteristic dimensions d_c, d_1 and d_f, calculated according to the Eqs. (47) and (52), is adduced in Fig. 11.21. As it follows from the plots of this figure, for PC the inequality (the Eq. (45)) is confirmed, which as a matter of fact is multifractal definition. The dependences $d_c(T)$ and $d_1(T)$ are similar and their absolute values are close, that is explained by the indicated above intercommunication of f_g and U change [89]. Let us note, that dimension d_1 controls only yield strain ε_Y and dimension d_c, both ε_Y and yield stress σ_Y. At approaching to glass transition temperature, that is, at $T \circledR T_g$, the values d_c, d_1 and d_f become approximately equal, that is, rubber is a regular fractal. Thus, with multifractal formalism positions the glass transition can be considered as the transition of structure from multifractal to the regular fractal. Additionally it is easy to show the fulfillment of the structure thermodynamically stability condition [90]:

$$d_f(d - d_1) = d_c(d - d_c). \tag{57}$$

Hence, the authors [73] considered the multifractal concept of amorphous glassy polymers yielding process. It is based on the hypothesis, that the yielding process presents itself structural transition from multifractal to regular fractal. In a amorphous glassy polymers Shennone informational entropy to an approximation of constant coincides with entropy, which is due to polymer fluctuation free volume change. The approximate quantitative estimations confirm the offered hypothesis correctness. The postulate about yielding process realization possibility only for polymers fractal structure, presented as a physical fractal within the definite range of linear scales, is the main key conclusion from the stated above results. What is more, the fulfilled estimations strengthen this definition—yielding is possible for polymer multifractal structure only and represents itself the structural transition multifractal—regular fractal [73].

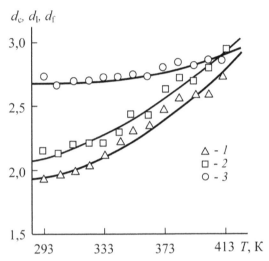

FIGURE 11.21   The temperature dependences of correlation d_c (1), informational d_I (2) and Hausdorff d_f (3) dimensions of PC multifractal structure [73].

Let us consider in the present chapter conclusion the treatment of dependences of yield stress on strain rate and crystalline phase structure for semicrystalline polymers [77]. As it known [91], the clusters relative fraction φ_{cl} is an order parameter of polymers structure in strict physical significance of this term and since the local order was postulated as having thermofluctuation origin, then φ_{cl} should be a function of testing temporal scale in virtue of super-position temperature time.

Having determined value t as duration of linear part of diagram load–time (P–t) in impact tests and accepting is equal to clusters relative fraction in quasistatic tensile tests [92], the values $\varphi_{cl}(t)$ can be estimated. In Fig. 11.22, the dependences $\varphi_{cl}(t)$ on strain rate for HDPE and polypropylene (PP) are shown, which demonstrate φ_{cl} increase at strain rate growth, that is, tests temporal scale decrease.

The elasticity modulus E value decreases at $\dot{\varepsilon}$ growth (Fig. 11.23). This effect cause (intermolecular bonds strong anharmonicity) is described in detail in Refs. [49, 94, 95]. In Fig. 11.24, the dependences of elasticity modulus E on structure fractal dimension d_f are adduced for HDPE and PP, which turned out to be linear and passing through coordinates origin. As it is known [84], the relation between E and d_f is given by the equation:

$$E = Gd_f,$$
(58)

where G is a shear modulus.

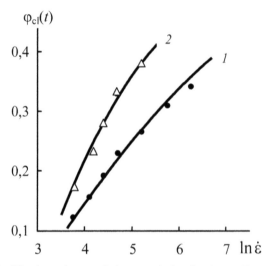

FIGURE 11.22 The dependences of clusters relative fraction $\varphi_{cl}(t)$ on strain rate $\dot{\varepsilon}$ in logarithmic coordinates for HDPE (1) and PP (2) [93].

From the Eq. (58) and Fig. 11.24 it follows, that for both HDPE and PP G is const. This fact is very important for further interpretation [77].

Let us estimate now determination methods of crystalline σ_Y^{cr} and non-crystalline σ_Y^{nc} regions contribution to yield stress σ_Y of semicrystalline polymers. The value σ_Y^{cr} can be determined as follows [96]:

$$\sigma_Y^{cr} = \frac{Gb}{2\pi}\left(\frac{K}{S}\right)^{1/2},$$
(59)

where b is Burgers vector, determined according to the equation (7), K is crystallinity degree, S is cross-sectional area, which is equal to 14.35 and 26.90 Å² for HDPE and PP, accordingly [97].

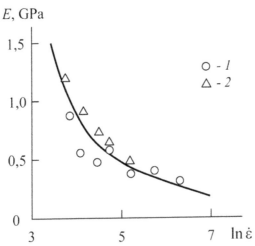

FIGURE 11.23   The dependence of elasticity modulus E on strain rate $\dot{\varepsilon}$ in logarithmic coordinates for HDPE (1) and PP (2) [93].

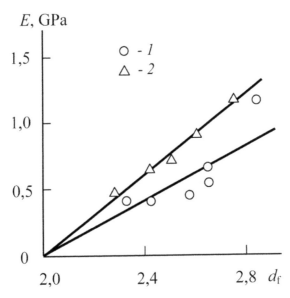

FIGURE 11.24   The dependences of elasticity modulus E on fractal dimension of structure d_f for HDPE (1) and PP (2) [93].

As one can see, all included in the Eq. (59) parameters for each polymer are constant, from which it follows, that σ_Y^{cr} =const. The value σ_Y^{cr} = 16.3 MPa for HDPE and 17.6 MPa, for PP. Thus, the change, namely, increase σ_Y at $\dot{\varepsilon}$ growth is defined by polymers noncrystalline regions contribution σ_Y^{nc} [77].

The Grist dislocation model [98] assumes the formation in polymers crystalline regions of screw dislocation (or such dislocation pair) with Burgers vector b and the yield process is realized at the formation of critical nucleus domain with size u^*:

$$u^* = \frac{Bb}{2\pi\tau_Y},\tag{60}$$

where B is an elastic constant, τ_Y shear yield stress.

In its turn, the domain with size U^* is formed at energetic barrier ΔG^* overcoming [98]:

$$\Delta G^* = \frac{Bl^2 l_d}{2\pi}\left[\ln\left(\frac{u^*}{r_0}\right)-1\right],\tag{61}$$

where l_d is dislocation length, which is equal to crystalline thickness, r_0 is dislocation core radius.

In the model [98], it has been assumed, that nucleus domain with size u^* is formed in defect-free part of semicrystalline polymer, that is, in crystallite. Within the frameworks of model [1] and in respect to these polymers amorphous phase structure such region is loosely packed matrix, surrounding a local order region (cluster), whose structure is close enough to defect-free polymer structure, postulated by the Flory "felt" model [16, 17]. In such treatment the value u^* can be determined as follows [43]:

$$u^* = R_{cl}-r_{cl},\tag{62}$$

where R_{cl} is one half of distance between neighboring clusters centers, r_{cl} is actually cluster radius.

The value R_{cl} is determined according to the equation [18]:

$$R_{cl} = 18\left(\frac{2v_e}{F}\right)^{-1/3}, \text{ Å},\tag{63}$$

where v_e is macromolecular binary hooking network density, values of which for HDPE and PP are adduced in Ref. [99], $F = 4$.

The value r_{cl} can be determined as follows [77]:

$$r_{cl} = \left(\frac{n_{st} S}{\pi \eta_{pac}} \right),$$

(64)

where n_{st} is statistical segments number per one cluster ($n_{st} = 12$ for HDPE and 15, for PP [77]), η_{pac} is packing coefficient, for the case of dense packing equal to 0.868 [100].

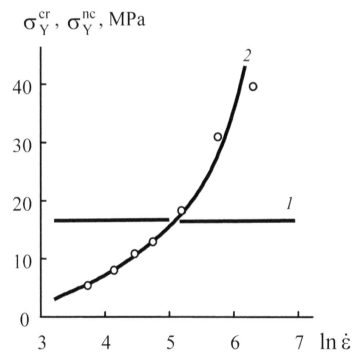

FIGURE 11.25 The dependences of crystalline σ_Y^{cr} (1) and noncrystalline σ_Y^{nc} (2) regions contributions in yield stress value σ_Y on strain rate in logarithmic coordinates for HDPE [93].

The shear modulus G, which, has been noted above, is independent on $\dot{\varepsilon}$, is accepted as B. Now the value σ_Y^{cr} can be determined according to the Eq. (60), assuming, that n_{st} changes proportionally to φ_{cl} (Fig. 11.22). In Fig. 11.25, the dependences of σ_Y^{cr} and σ_Y^{nc} on strain rate $\dot{\varepsilon}$ are adduced for HDPE (the similar picture was obtained for PP). As it follows from the data of this figure, at the condition σ_Y^{cr} = const the value σ_Y^{nc} grows at $\dot{\varepsilon}$ increase and at $\dot{\varepsilon} \approx 150 \text{ s}^{-1}$ $\sigma_Y^{cr} > \sigma_Y^{nc}$, that is, a noncrystalline regions contribution in σ_Y begins to prevail [93].

In Fig. 11.26, the comparison of the experimental σ_Y and calculated theoretically σ_Y^T as sum ($\sigma_Y^{cr} + \sigma_Y^{nc}$) dependences of yield stress on $\dot{\varepsilon}$ for HDPE and PP are adduced. As one can see, a good correspondence of theory and experiment is obtained, confirming the offered above model of yielding process for semicrystalline polymers.

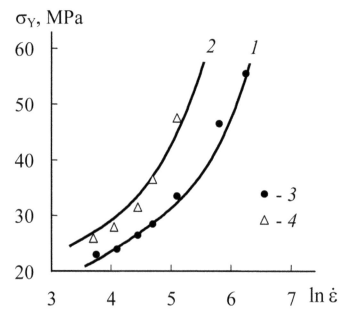

FIGURE 11.26 Comparison of the experimental (1, 2) and theoretical (3, 4) dependences of yield stress σ_Y on strain rate $\dot{\varepsilon}$ in logarithmic coordinates for HDPE (1, 3) and PP (2, 4) [93]

Let us note in conclusion the following. As it follows from the comparison of the data of Figs. 11.23, 11.24 and 11.26, σ_Y increase occurs at E

reduction and G constancy, that contradicts to the assumed earlier σ_Y and E proportionality [101]. The postulated in Ref. [101] σ_Y and E proportionality is only an individual case, which is valid either at invariable structural state or at the indicated state, changing by definite monotonous mode [77].

Hence, the stated above results demonstrate nonzero contribution of noncrystalline regions in yield stress even for such semicrystalline polymers, which have devitrificated amorphous phase in testing conditions. At definite conditions noncrystalline regions contribution can be prevailed. Polymers yield stress and elastic constants proportionality is not a general rule and is fulfilled only at definite conditions.

The authors [102] use the considered above model [93] for branched polyethylenes (BPE) yielding process description. As it is known [103], the crystallinity degree K, determined by samples density, can be expressed as follows:

$$K = \alpha_c + \alpha_{if}, \qquad (65)$$

where α_c and α_{if} are chains units fractions in perfect crystallites and anisotropic interfacial regions, accordingly.

The Eq. (59) with replacement of K by α_c was used for the value σ_Y^{cr} estimation. Such estimations have shown that the value σ_Y^{cr} is always smaller than macroscopic yield stress σ_Y. In Fig. 11.27, the dependence of crystalline phase relative contribution in yield stress σ_Y^{cr}/σ_Y on α_c is adduced. This dependence is linear and at $\alpha_c = 0$ the trivial result $\sigma_Y^{cr} = 0$ is obtained. Let us note that this extrapolation assumes $\sigma_Y \neq 0$ at $\alpha_c = 0$. At large α_c the crystalline phase contribution is prevailed and at $\alpha_c = 0{,}75$ $\sigma_Y^{cr}/\sigma_Y = 1$ (Fig. 11.27).

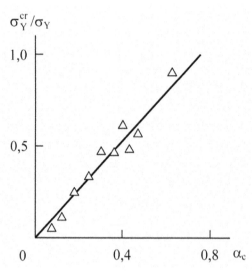

FIGURE 11.27 The dependence of crystalline regions relative contribution σ_Y^{cr}/σ_Y in yield stress on perfect crystallites fraction α_c for series of BPE [102]

The dependence $\alpha_{if}(\alpha_c)$ showed α_{if} linear reduction at α_c growth. Such α_{if} change and simultaneous σ_Y^{cr}/σ_Y increasing (Fig. 11.27) at α_c growth assume local order degree reduction, determining noncrystalline regions contribution in σ_Y, at crystallinity degree enhancement. Besides, the correlation $\alpha_{if}(\alpha_c)$ shows that local order regions of BPE noncrystalline regions are concentrated mainly in anisotropic interfacial regions, as it has been assumed earlier [18]. At $\alpha_c = 0.82$ $\alpha_{if} = 0$, that corresponds to the data of Fig. 11.27.

The common fraction of the ordered regions (clusters and crystallites) φ_{ord} can be determined according to the percolation relationship [91]:

$$\varphi_{ord} = 0.03(T_m - T)^{0.55},$$ (66)

where T_m and T are melting and testing temperatures, respectively.

Then clusters relative fraction φ_{cl} is estimated from the obvious relationship [102]:

$$\varphi_{cl} = \varphi_{ord} - \alpha_c.$$ (67)

The comparison of the experimental and estimated as a sum of crystalline and noncrystalline regions contributions in yield stress theoretical values α_Y shows their good correspondence for considered BPE.

It has been noted earlier [104, 105], that stress decay beyond stress ("yield tooth") for polyethylenes is expressed the stronger the greater value α_c is. For amorphous polymers it has been shown that the indicated "yield tooth" is due to instable clusters decay in yielding process and this decay is expressed the clearer the higher instable clusters relative fraction [39]. By analogy with the indicated mechanism the authors [102] assume that "yield tooth" will be the stronger the larger crystallites fraction is subjected to mechanical disordering (partial melting) in yielding process. The indicated fraction of crystallites χ_{cr} is determined as difference [102]:

$$\chi_{cr} = \chi - \alpha_{am}, \tag{68}$$

where χ is polymer fraction, subjecting to elastic deformation, α_{am} is fraction of amorphous phase.

The value χ can be determined within the frameworks of polymers plasticity fractal concept [35] according to the Eq. (9). The dependence of χ_{cr} on α_c shows χ_{cr} increasing at α_c growth [102]. At small α_c values all crystallites are subjected to disordering owing to that "yield tooth" in BPE curves stress-strain is absent and these curves are acquired the form, which is typical for rubbers. Hence, stress decay beyond yield stress intensification is due to χ_{cr} growth [103].

REFERENCES

1. Kozlov, G. V., Belousov, V. N., Serdyuk, V. D., Kuznetsov, E. N. Defects of Polymers Amorphous State Structure. Fizika i Technika Yysokikh Davlenii, 1995, v. 5, № 3, p. 59–64.
2. Kozlov, G. V., Beloshenko, V. A., Varyukhin, V. N. Evolution of Dissipative Structures in Yielding Process of Cross-Linked Polymers. Prikladnaya Mekhanika i Tekhnicheskaya Fizika, 1996, v. 37, № 3, p. 115–119.
3. Honeycombe, R. W. K. The Plastic Deformation of Metals. London, Edward Arnold Publishers, Ltd, 1968, 398 p.
4. Argon, A. S. A theory for the Low-Temperature Plastic Deformation of Glassy Polymers. Phil. Mag., 1974, v 29, 1988, № 1, p. 149–167.

5. Argon, A. S. Physical Basis of Distortional and Dilational Plastic Flow in Glassy Polymers. J. Macromol. Sci.-Phys., 1973, v. 88, № 3–4, p. 573–596.
6. Bowden, P. B., Raha, S. A. Molecular Model for Yield and Flow in Amorphous Glassy Polymers Making use of a Dislocation Analoque. Phil. Mag., 1974, v. 29, № 1, p. 149–165.
7. Escaig, B. The Physics of Plastic Behaviour of Crystalline and Amorphous Solids. Ann. Phys., 1978, v. 3, № 2, p. 207–220.
8. Pechhold, W. R., Stoll, B. Motion of Segment Dislocations as a Model for Glass Relaxation. Polymer Bull., 1982, v. 7, № 4, p. 413–416.
9. Sinani, A. B., Stepanov, V. A. Prediction of Glassy Polymer Deformation Properties with the Aid of Dislocation Analogues. Mekhanika Kompozitnykh Materialov, 1981, v. 17, № 1, p. 109–115.
10. Oleinik, E. F., Rudnev, S. M., Salamatina, O. B., Nazarenko, S. I., Grigoryan, G. A. Two Modes of Glassy Polymers Plastic Deformation. Doklady AN SSSR, 1986, v. 286, № 1. p. 135–138.
11. Melot, D., Escaig, B., Lefebvre, J. M., Eustache, R. R., Laupretre, F. Mechanical Properties of Unsaturated Polyester Resins in Relation to their Chemical Strcucture: 2. Plastic Deformation Behavior. J. Polymer Sci.: Part B: Polymer Phys., 1994, v. 32, № 11, p. 1805–1811.
12. Boyer, R. F. General Reflections on the Symposium on Physical Structure of the Amorphous State. J. Macromol. Sci., Phys., 1976, v. B12, № 12, № 2, p. 253–301.
13. Fischer, E. W., Dettenmaier, M. Structure of Polymer Glasses and Melts. J. Non-Cryst. Solids, 1978, v. 31, № 1–2, p. 181–205.
14. Wendorff, J. H. The Structure of Amorphous Polymers. Polymer, 1982, v. 23, № 4, p. 543–557.
15. Nikol skii, V. G., Plate, I. V., Fazlyev, F. A., Fedorova, E. A., Filippov, V. V., Yudaeva, L. V. Structure of Polyolefins Thin Films, Obtained by Quenching of Melt up to 77K. Vysokomolek. Soed. A, 1983, v. 25, № 11, p. 2366–2371.
16. Flory, P. J. Conformations of Macromolecules in Condenced Phases. Pure Appl. Chem., 1984, v. 56, № 3, p. 305–312.
17. Flory, P. J. Spatial Configuration of Macromolecular Chains. Brit. Polymer, J., 1976, v. 8, № 1–10.
18. Kozlov, G. V., Zaikov, G. E. Structure of the Polymer Amorphous State. Utrecht, Boston, Brill Academic Publishers, 2004, 465 p.
19. Kozlov, G. V., Shogenov, V. N., Mikitaev, A. K. A Free Volume Role in Amorphous Polymers Forced Elasticity Process. Doklady AN SSSR, 1988, v. 298, № 1, p. 142–144.
20. Kozlov, G. V., Afaunova, Z. I., Zaikov, G. E. The Theoretical Estimation of Polymers Yield Stress. Electronic Zhurnal "Issledovano v Rossii", 98, p. 1071–1080, 2002. http:.zhurnal. ape. relarn. ru/articles /2002/ 098. pdf.
21. Liu, R. S., Li, J. Y. On the Structural Defects and Microscopic Mechanism of the High Strength of Amorphous Alloys. Mater. Sci. Engng., 1989, v. A 114, № 1, p. 127–132.
22. Alexanyan, G. G., Berlin, A. A., Gol'danskii, A. V., Grineva, N. S., Onitshuk, V. A., Shantarovich, V. P., Safonov, G. P. Study by the Positrons Annihilation Method of Annealing and Plastic Deformation Influence on Polyarylate Microstructure. Khimicheskaya Fizika, 1986, v. 5, № 9, p. 1225–1234.

23. Balankin, A. S., Bugrimov, A. L., Kozlov, G. V., Mikitaev, A. K., Sanditov, D. S. The Fractal Structure and Physical-Mechanical Properties of Amorphous Glassy Polymers. Doklady AN, 1992, v, 326, № 3, p. 463–466.

24. Kozlov, G. V., Novikov, V. U. The Cluster Model of Polymers Amorphous State. Uspekhi Fizicheskikh Nauk, 2001, v. 171, № 7, p. 717–764.

25. McClintock, F. A., Argon, A. S. Mechanical Behavior of Materials. Massachusetts, Addison-Wesley Publishing Company, Inc., 1966, 432 p.

26. Wu, S. Secondary Relaxation, Brittle-Ductile Transition Temperature, and Chain Structure. J. Appl. Polymer Sci., 1992, v. 46, № 4, p. 619–624.

27. Sanditov, D. S., Bartenev, G. M. Physical Properties of Disordered Structures. Novosibirsk, Nauka, 1982, 256 p.

28. Mil man, L. D., Kozlov, G. V. Polycarbonate Nonelastic Deformation Simulation in Impact Loading Conditions with the Aid of Dislocations Analogues. In book: Polycondencation Processes and Polymers, Nal'chik, KBSU, 1986, p. 130–141.

29. Shogenov, V. N., Kozlov, G. V., Mikitaev, A. K. Prediction of Forsed Elasticity of Rigid-Chain Polymers. Vysokomolek. Soed. A, 1989, v. 31, № 8, p. 1766–1770.

30. Sanditov, D. S., Kozlov, G. V. About Correlation Nature between Elastic Moduli and Glass Transition Temperature of Amorphous Polymers. Fizika i Khimiya Stekla, 1993, v. 19, № 4, p. 593–601.

31. Peschanskaya, N. N., Bershtein, V. A., Stepanov, V. A. The Connection of Glassy Polymers Creep Activation Energy with Cohesion Energy. Fizika Tverdogo Tela, 1978, v. 20, № 11, p. 3371–3374.

32. Shogenov, V. N., Belousov, V. N., Potapov, V. V., Kozlov, G. V., Prut, E. V. The Glassy Polyarylatesulfone Curves stress-strain Description within the Frameworks of High-Elasticity Concepts. Vysokomolek. Soed. A, 1991, v. 33, № 1, p. 155–160.

33. Mashukov, N. I., Gladyshev, G. P., Kozlov, G. V. Structure and Properties of High Density Polyethylene Modified by High-Disperse Mixture Fe and FeO. Vysokomolek. Soed. A, 1991, v. 33, № 12, p. 2538–2546.

34. Beloshenko, V. A., Kozlov, G. V. The Cluster Model Application for Epoxy Polymers Yielding Process. Mechanika Kompozitnykh Materialov, 1994, v. 30, № 4, p. 451–454.

35. Balankin, A. S., Bugrimov, A. L. The Fractal Theory of Polymers Plasticity. Vysokomolek. Soed. A, 1992, v. 34, № 10, p. 135–139.

36. Andrianova, G. P., Kargin, V. A. To Necking Theory at Polymers Tension. Vysokomolek. Soed. A, 1970, v. 12, № 1, p. 3–8.

37. Kachanova, I. M., Roitburd, A. L. Plastic Deformation Effect on New Phase Inclusion Equilibrium Form and Thermodynamical Hysteresis. Fizika Tverdogo Tela, 1989, v. 31, № 4, p. 1–9.

38. Hartmann, B., Lee, G. F., Cole, R. F. Tensile Yield in Polyethylene. Polymer Engng. Sci., 1986, v. 26, № 8, p. 554–559.

39. Kozlov, G. V., Beloshenko, V. A., Gazaev, M. A., Novikov, V. U. Mechanisms of Yielding and Forced High-Elasticity of Cross-Linked Polymers. Mekhanika Kompozitnykh Materialov, 1996, v. 32, № 2, p. 270–278.

40. Balankin, A. S. Synergetics of Deformable Body. Moscow, Publishers of Ministry of Defence SSSR, 1991, 404 p.

41. Filyanov, E. M. The Connection of Activation Parameters of Cross-Linked Polymers Deformation in Transient Region with Glassy State Properties Vysokomolek. Soed. A, 1987, v. 29, № 5, p. 975–981.
42. Kozlov, G. V., Sanditov, D. S. The Activation Parameters of Glassy Polymers Deformation in Impact Loading Conditions. Vysokomolek. Soed. B, 1992, v. 34, № 11, p. 67–72.
43. Kozlov, G. V., Shustov, G. B., Zaikov, G. E., Burmistr, M. V., Korenyako, V. A. Polymers Yielding Description within the Franeworks of Thermodynamical Hierarchical Model. Voprosy Khimii I Khimicheskoi Tekhnologii, 2003, № 1, p. 68–72.
44. Gladyshev, G. P., Thermodynamics and Macrokinetics of Natural Hierarchical Processes. Moscow, Nauka, 1988, 290 p.
45. Gladyshev, G. P., Gladyshev, D. P. The Approximate Thermodynamical Equation for Nonequilibrium Phase Transitions. Zhurnal Fizicheskoi Khimii, 1994, v. 68, № 5, p. 790–792.
46. Kozlov, G. V., Zaikov, G. E. Thermodynamics of Polymer Structure Formation in an Amorphous State. In book: Fractal Analysis of Polymers: From Synthesis to Composites Ed. Kozlov, G. V., Zaikov, G. E., Novikov, V. U. New York, Nova Science Publishers, Inc, 2003, p. 89–97.
47. Matsuoka, S., Bair, H. E. The Temperature Drop in Glassy Polymers during Deformation. J. Appl. Phys., 1977, v. 48, № 10, p. 4058–4062.
48. Kozlov, G. V., Zaikov, G. E. Formation Mechanisms of Local Order in Polymers Amorphous State Structure. Izvestiya KBNC RAN, 2003, № 1(9), p. 54–57.
49. Kozlov, G. V., Sanditov, D. S. Anharmonic Effects and Physical-Mechanical Properties of Polymers. Novosibirsk, Nauka, 1994, 261 p.
50. Matsuoka, S., Aloisio, C. Y., Bair, H. E. Interpretation of Shift of Relaxation Time with Deformation in Glassy Polymers in Terms of Excess Enthalpy. J. Appl. Phys., 1973, v. 44, № 10, p. 4265–4268.
51. Kosa, P. N., Serdyuk, V. D., Kozlov, G. V., Sanditov, D. S. The Comparative Analysis of Forced Elasticity Process for Amorphous and Semicrystalline Polymers. Fizika I Tekhnika Vysokikh Davlenii, 1995, v. 5, № 4, p. 70–81.
52. Kargin, V. A., Sogolova, I. I. The Studu of Crystalline Polymers Mechanical Properties. I. Polyamides Zhurnal Fizicheskoi Khimii, 1953, v. 27, № 7, p. 1039–1049.
53. Gent, A. N., Madan, S. Plastic Yielding of Partially Crystalline Polymers. J. Polymer Sci." Part B: Polymer Phys., 1989, v. 27, № 7, p. 1529–1542.
54. Mashukov, N. I., Belousov, V. N., Kozlov, G. V., Ovcharenko, E. N., Gladychev, G. P. The Connection of Forced Elasticity Stress and Structure for Semicrystalline Polymers. Izvestiya AN SSSR, seriya khimicheskaya, 1990, № 9, p. 2143–2146.
55. Sanditov, D. S., Kozlov, G. V., Belousov, V. N., Lipatov Yu.S. The Cluster Model and Fluctuation Free Volume Model of Polymeric Glasses. Fizika i Khimiya Stekla, 1994, v. 20, № 1, p. 3–13.
56. Graessley, W. W., Edwards, S. F. Entanglement Interactions in Polymers and the Chain Contour Concentration. Polymer, 1981, v. 22, № 10, p. 1329–1334.
57. Adams, G. W., Farris, R. Y., Latent Energy of Deformation of Amorphous Polymers. 1. Deformation Calorimetry. Polymer, 1989, v. 30, № 9, p. 1824–1828.
58. Oldham, K., Spanier, J. Fractional Calculus. London, New York, Academic Press, 1973, 412 p.

59. Samko, S. G., Kilbas, A. A., Marishev, O. I. Integrals and Derivatives of Fractional Order and their some Applications. Minsk, Nauka i Tekhnika, 1987, 688 p.
60. Nigmatullin, R. R. Fractional Integral and its Physical Interpretation. Teoreticheskaya i Matematicheskaya Fizika, 1992, v. 90, № 3, p. 354–367.
61. Feder, F. Fractals. New York, Plenum Press, 1989, 248 p.
62. Kozlov, G. V., Shustov, G. B., Zaikov, G. E. Polymer Melt Structure Role in Heterochain Polyeters Thermooxidative Degradation Process. Zhurnal Prikladnoi Khimii, 2002, v. 75, № 3, p. 485–487.
63. Kozlov, G. V., Batyrova, H. M., Zaikov, G. E. The Structural Treatment of a number of Effective Centres of Polymeric Chains in the Process of the Thermooxidative Degradation. J. Appl. Polymer Sci., 2003, v. 89, № 7, p. 1764–1767.
64. Kozlov, G. V., Sanditov, D. S., Ovcharenko, E. N. Plastic Deformation Energy and Structure of Amorphous Glassy Polymers. Proceeding of Internat. Interdisciplinary Seminar "Fractals and Applied Synergetics FiAS-01", 26–30 November 2001, Moscow, p. 81–83.
65. Meilanov, R. P., Sveshnikova, D. A., Shabanov, O. M. Sorption Kinetics in Systems with Fractal Structure. Izvestiya VUZov, Severo-Kavkazsk. Region, estestv. nauki, 2001, № 1, p. 63–66.
66. Kozlov, G. V., Mikitaev, A. K. The Fractal Analysis of Yielding and Forced High-Elasticity Processes of Amorphous Glassy Polymers. Mater. I-th All-Russian Sci-Techn. Conf. "Nanostructures in Polymers and Polymer Nanocomposites." Nal chik, KBSU, 2007, p. 81–86.
67. Kekharsaeva, E. R., Mikitaev, A. K., Aleroev, T. S. Model of Stress-Strain Characteristics of Chlor-Containing Polyesters on the Basis of Derivatives of Fractional Order. Plast. Massy, 2001, № 3, p. 35.
68. Novikov, V. U., Kozlov, G. V. Structure and Properties of the Polymers within the Frameworks of Fractal Approach. Uspekhi Khimii, 2000, v. 69, № 6, p. 572–599.
69. Shogenov, V. N., Kozlov, G. V., Mikitaev, A. K. Prediction of Mechanical Behavior, Structure and Properties of Film polymer Samples at Quasistatic Tension. In book: Polycondensation Reactions and Polymers. Selected Works. Nal chik, KBSU, 2007, p. 252–270.
70. Lur e, E. G., Kovriga, V. V. To Question about Unity of Deformation Mechanism and Spontaneous Lengthening of Rigid-Chain Polymers. Mekhanika Polimerov, 1977, № 4, p. 587–593.
71. Lur e, E. G., Kazaryan, L. G., Kovriga, V. V., Uchastkina, E. L., Lebedinskaya, M. L., Dobrokhotova, M. L., Emel yanova, L. M. The Features of Crystallization and Deformation of Polyimide Film PM. Plast. Massy., 1970, № 8, p. 59–63.
72. Shogenov, V. N., Kozlov, G. V., Mikitaev, A. K. Prediction of Rigid-Chain Polymers Mechanical Properties in Elasticity Region. Vysokomolek. Soed. B, 1989, v. 31. № 7, p. 553–557.
73. Kozlov, G. V., Yanovskii Yu.G., Karnet Yu.N. The Generalized Fractal Model of Amorphous Glassy Polymers Yielding Process. Mekhanika Kompozitsionnykh Materialov i Konstruktsii, 2008, v. 14, № 2, p. 174–187.
74. Kopelman, R. Excitons Dynamics Mentioned Fractal one: Geometrical and Energetic Disorder. In book. Fractals in Physics. Ed. Pietronero, L., Tosatti, E. Amsterdam, Oxford, New York, Tokyo, North-Holland, 1986, p. 524–527.

75. Korshak, V. V., Vinogradova, S. V. Nonequilibrium Polycondensation. Moscow, Nauka, 1972, 696 p.
76. Vatrushin, V. E., Dubinov, A. E., Selemir, V. D., Stepanov, N. V. The Analysis of SHF Apparatus Complexity with Virtual Cathode as Dynamical Objects. In book: Fractals in Applied Physics. Ed. Dubinov, A. E. Arzamas-16, VNIIEF, 1995, p. 47–58.
77. Mikitaev, A. K., Kozlov, G. V. Fractal Mechanics of Polymer Materials. Nal chik, Publishers KBSU, 2008, 312 p.
78. Kozlov, G. V., Belousov, V. N., Mikitaev, A. K. Description of Solid Polymers as Quasitwophase Bodies. Fizika i Tekhika Vysokikh Davlenii, 1998, v. 8, № 1, p. 101–107.
79. Aloev, V. Z., Kozlov, G. V., Beloshenko, V. A. Description of Extruded Componors Structure and Properties within the Frameworks of Fractal Analysis. Izvestiya VUZov, Severo-Kavkazsk. Region, estesv. nauki, 2001, № 1, p. 53–56.
80. Argon, A. S., Bessonov, M. I. Plastic Deformation in Polyimides, with New Implications on the Theory of Plastic Deformation of Glassy Polymers. Phil. Mag., 1977, v. 35, № 4, p. 917–933.
81. Beloshenko, V. A., Kozlov, G. V., Slobodina, V. G., Prut, E. U., Grinev, V. G. Thermal Shrinkage of Extrudates of Ultra-High-Molecular Polyethylene and Polymerization-Filled Compositions on its Basis. Vysokomolek. Soed. B, 1995, v. 37, № 6, p. 1089–1092.
82. Balankin, A. S., Izotov, A. D., Lazarev, V. B. Synergetics and Fractal Thermomechanics of Inorganic Materials. I. Thermomechanics of Multifractals. Neorganicheskie Materialy, 1993, v. 29, № 4, p. 451–457.
83. Haward, R. N. Strain Hardening of Thermoplastics. Macromolecules, 1993, v. 26, № 22, p. 5860–5869.
84. Balankin, A. S. Elastic Properties of Fractals and Dynamics of Solids Brittle Fracture. Fizika Tverdogo Tela, 1992, v. 34, № 36 p. 1245–1258.
85. Kozlov, G. V., Afaunova, Z. I., Zaikov, G. E. Experimental Estimation of Multifractal Characteristics of Free Volume for Poly(vinyl acetate). Oxidation Commun., 2005, v. 28, № 4, p. 856–862.
86. Hentschel, H. G. E., Procaccia, I. The Infinite Number of Generalized Dimensions of Fractals and Strange Attractors Phys. D., 1983, v. 8, № 3, p. 435–445.
87. Williford, R. E. Multifractal Fracture. Scripta Metallurgica, 1988, v. 22, № 11, p. 1749–1754.
88. Kalinchev, E. L., Sakovtseva, M. B. Properties and Processing of Thermoplastics. Leningrad, Khimiya, 1983, 288 p.
89. Sanditov, D. S., Sangadiev, S.Sh. About Internal Pressure and Microhardness of Inorganic Glasses. Fizika i Khimiya Stekla, 1988, v. 24, № 6, p. 741–751.
90. Balankin, A. S. The Theoty of Elasticity and Entropic High-Elasticity of Fractals. Pis ma v ZhTF, 1991, v, 17, № 17, p. 68–72.
91. Kozlov, G. V., Gazaev, M. A., Novikov, V. U., Mikitaev, A. K. Simulation of Amorphous Polymers Structure as Percolation cluster. Ris ma v ZhTF, 1996, v. 22, № 16, p. 31–38.
92. Serdyuk, V. D., Kosa, P. N., Kozlov, G. V. Simulation of Polymers Forced Elasticity Process with the Aid of Dislocation Analoques. Fizika i Tekhnika Vysokikh Davlenii, 1995, v. 5, № 3, p. 37–42.

93. Kozlov, G. V. The Dependence of Yield Stress on Strain Rate for Semicrystalline Polymers. Manuscript disposed to V I N I I I RAN, Moscow, November 1, 2002, № 1884-B2002.

94. Kozlov, G. V., Shetov, R. A., Mikitaev, A. K.. Methods of Elasticity Modulus Measurement in Polymers Impact Tests. Vysokomolek. Soed. A, 1987, v. 29, № 5, p. 1109–1110.

95. Sanditov, D. S., Kozlov, G. V. Anharmonicity of Interatomic and Intermolecular Bonds and Physical-Mechanical Properties of Polymers. Fizika i Khimiya Stekla, 1995, v. 21, № 6, p. 547–576.

96. Belousov, V. N., Kozlov, G. V., Mashukov, N. I., Lipatov Yu.S. The Application of Dislocation Analogues for Yieldin Process Description in Crystallizable Polymers. Doklady AN, 1993, v. 328, № 6, p. 706–708.

97. Aharoni, S. M. Correlations between Chain Parameters and Failure Characteristics of Polymers below their Glass Transition Temperature. Macromolecules, 1985, v. 18, № 12, p. 2624–2630.

98. Crist, B. Yielding of Semicrystalline Polyethylene: a Quantitive Dislocation Model. Polymer Commun., 1989, v. 30, № 3, p. 69–71.

99. Wu, S. Chain Structure and Entanglement. J. Polymer Sci.: Part B: Polymer Phys., 1989, v. 27, № 4, p. 723–741.

100. Kozlov, G. V., Beloshenko, V. A., Varyukhin, V. N. Simulation of Gross-Linked Polymers Structure as Diffusion-Limited Aggregate. Ukrainskii Fizicheskii Zhurnal, 1988, v. 43, № 3, p. 322–323.

101. Brown, N. The relationship between Yield Point and Modulus for Glassy Polymers. Mater. Sci. Engng., 1971, v. 8, № 1, p. 69–73.

102. Kozlov, G. V., Shustov, G. B., Aloev, V. Z., Ovcharenko, E. N. Yielding Process of Branched Polyethylenes. Proceedings of II All-Russian Sci-Pract. Conf. "Innovations in Mechanical Engineering." Penza, PSU, 2002, p. 21–23.

103. Mandelkern, L., The Relation between Structure and Properties of Crystalline Polymers. Polymer, J., 1985, v. 17, № 1, p. 337–350.

104. Peacock, A. J., Mandelkern, L. The Mechanical Properties of Random Copolymers of Ethylene: Force-Elongation Relations. J. Polymer Sci.: Part B: Polymer Phys., 1990, v. 28, № 11, p. 1917–1941.

105. Kennedy, M. A., Reacock, A. J., Mandelkern, L. Tensile Properties of Crystalline Polymers: Linear Polyethylene. Macromolecules, 1994, v. 27, № 19, p. 5297–5310.

CHAPTER 12

LECTURE NOTES ON QUANTUM CHEMICAL CALCULATION

V. A. BABKIN, G. E. ZAIKOV, D. S. ANDREEV,
YU. KALASHNIKOVA, YU. S. ARTEMOVA, and D. V. SIVOVOLOV

CONTENTS

12.1   QUANTUM-CHEMICAL CALCULATION OF MOLECULE 1,4-(1,1'-3,3'-DIINDENYL)BUTANE BY METHOD AB INITIO

12.1.1   INTRODUCTION

In this section, for the first time quantum chemical calculation of a molecule of 1,4-(1,1'-diindenyl)butane is executed by method AB INITIO with optimization of geometry on all parameters. The optimized geometrical and electronic structure of this compound is received. Acid power of 1,4-(1,1'-diindenyl)butane is theoretically appreciated. It is established, than it to relate to a class of very weak H-acids (pKa = +32, where pKa-universal index of acidity).

12.1.2   AIMS AND BACKGROUNDS

The aim of this work is a study of electronic structure of molecule 1,4-(1,1'-3,3'-diindenyl)butane [1] and theoretical estimation its acid power by quantum-chemical method AB INITIO in base 6–311G**. The calculation was done with optimization of all parameters by standard gradient method built-in in PC GAMESS [2]. The calculation was executed in approach the insulated molecule in gas phase. Program MacMolPlt was used for visual presentation of the model of the molecule [3].

12.1.3   METHODOLOGY

Geometric and electronic structures, general and electronic energies of molecule 1,4-(1,1'-3,3'-diindenyl)butane was received by method AB INITIO in base 6–311G** and are shown on Fig. 12.1 and in Table 12.1. The universal factor of acidity was calculated by formula: pKa = $49.04 - 134.6 * q_{max}^{H+}$ [4, 5] (where, q_{max}^{H+} – a maximum positive charge on atom of the hydrogen $q_{max}^{H+} = +0.13$ (for 1,4-(1,1'-3,3'-diindenyl)butane q_{max}^{H+} alike Table 12.1)). This same formula is used in Refs. [6–16]. pKa=32.

Quantum-chemical calculation of molecule 1,4-(1,1'-3,3'-diindenyl) butane by method AB INITIO in base 6–311G** was executed for the first time. Optimized geometric and electronic structure of this compound was received. Acid power of molecule 1,4-(1,1'-3,3'-diindenyl)butane was theoretically evaluated (pKa=32). This compound pertains to class of very weak H-acids (pKa>14).

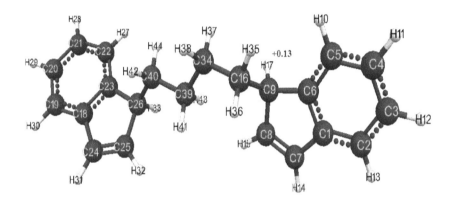

FIGURE 12.1 Geometric and electronic molecule structure of 1,4-(1,1'-3,3'-diindenyl) butane (E_0= −2,220,960 kDg/mol, E_{el}= −6,268,159 kDg/mol).

TABLE 12.1 Optimized bond lengths, valence corners and charges on atoms of the molecule 1,4-(1,1'-3,3'-diindenyl)butane.

Bond lengths	R, A	Valence corners	Grad	Atom	Charges on atoms
C(2)-C(1)	1.38	C(5)-C(6)-C(1)	120	C(1)	−0.05
C(3)-C(2)	1.39	C(1)-C(2)-C(3)	119	C(2)	−0.07
C(4)-C(3)	1.39	C(2)-C(3)-C(4)	121	C(3)	−0.09
C(5)-C(4)	1.39	C(3)-C(4)-C(5)	121	C(4)	−0.09
C(6)-C(5)	1.38	C(4)-C(5)-C(6)	119	C(5)	−0.06
C(6)-C(1)	1.39	C(2)-C(1)-C(6)	121	C(6)	−0.08
C(7)-C(1)	1.47	C(2)-C(1)-C(7)	131	C(7)	−0.06
C(8)-C(7)	1.33	C(9)-C(8)-C(7)	112	C(8)	−0.10
C(8)-C(9)	1.52	C(1)-C(7)-C(8)	110	C(9)	−0.12
C(9)-C(6)	1.52	C(6)-C(9)-C(8)	101	H(10)	+0.08

TABLE 12.2　*(Continued)*

Bond lengths	R, A	Valence corners	Grad	Atom	Charges on atoms
H(10)-C(5)	1.08	C(16)-C(9)-C(8)	120	H(11)	+0.09
H(11)-C(4)	1.08	C(5)-C(6)-C(9)	121	H(12)	+0.09
H(12)-C(3)	1.08	C(1)-C(6)-C(9)	124	H(13)	+0.08
H(13)-C(2)	1.08	C(4)-C(5)-H(10)	125	H(14)	+0.09
H(14)-C(7)	1.07	C(3)-C(4)-H(11)	123	H(15)	+0.10
H(15)-C(8)	1.07	C(2)-C(3)-H(12)	114	C(16)	−0.16
C(16)-C(9)	1.54	C(1)-C(2)-H(13)	109	**H(17)**	**+0.13**
H(17)-C(9)	1.09	C(1)-C(7)-H(14)	120	C(18)	−0.05
C(18)-C(23)	1.40	C(7)-C(8)-H(15)	109	C(19)	−0.07
C(19)-C(18)	1.38	C(9)-C(8)-H(15)	121	C(20)	−0.09
C(20)-C(19)	1.39	C(6)-C(9)-C(16)	131	C(21)	−0.09
C(21)-C(20)	1.39	C(6)-C(9)-H(17)	119	C(22)	−0.07
C(22)-C(21)	1.39	C(22)-C(23)-C(18)	121	C(23)	−0.08
C(23)-C(22)	1.38	C(26)-C(23)-C(18)	121	C(24)	−0.05
C(23)-C(26)	1.52	C(23)-C(18)-C(19)	130	C(25)	−0.10
C(24)-C(18)	1.47	C(24)-C(18)-C(19)	119	C(26)	−0.12
C(25)-C(24)	1.33	C(18)-C(19)-C(20)	101	H(27)	+0.08
C(26)-C(25)	1.52	C(19)-C(20)-C(21)	114	H(28)	+0.09
C(26)-C(40)	1.54	C(20)-C(21)-C(22)	108	H(29)	+0.09
H(27)-C(22)	1.08	C(26)-C(23)-C(22)	110	H(30)	+0.09
H(28)-C(21)	1.08	C(21)-C(22)-C(23)	114	H(31)	+0.09
H(29)-C(20)	1.08	C(25)-C(26)-C(23)	112	H(32)	+0.10
H(30)-C(19)	1.08	C(40)-C(26)-C(23)	114	H(33)	+0.13
H(31)-C(24)	1.07	C(23)-C(18)-C(24)	120	C(34)	−0.18
H(32)-C(25)	1.07	C(18)-C(24)-C(25)	120	H(35)	+0.10
H(33)-C(26)	1.09	C(40)-C(26)-C(25)	120	H(36)	+0.10
C(34)-C(16)	1.53	C(24)-C(25)-C(26)	121	H(37)	+0.09
H(35)-C(16)	1.09	C(39)-C(40)-C(26)	124	H(38)	+0.10
H(36)-C(16)	1.09	C(21)-C(22)-H(27)	125	C(39)	−0.19
H(37)-C(34)	1.09	C(20)-C(21)-H(28)	109	C(40)	−0.15
H(38)-C(34)	1.09	C(19)-C(20)-H(29)	108	H(41)	+0.10
C(39)-C(34)	1.53	C(18)-C(19)-H(30)	116	H(42)	+0.11
C(40)-C(39)	1.53	C(18)-C(24)-H(31)	108	H(43)	+0.09
H(41)-C(39)	1.09	C(24)-C(25)-H(32)	109	H(44)	+0.10
H(42)-C(40)	1.09	C(25)-C(26)-H(33)	109		
H(43)-C(39)	1.09	C(40)-C(26)-H(33)	107		
H(44)-C(40)	1.09	C(9)-C(16)-C(34)	117		
		C(9)-C(16)-H(35)	112		
		C(9)-C(16)-H(36)	109		
		C(16)-C(34)-H(37)	110		
		C(16)-C(34)-H(38)	110		
		C(16)-C(34)-C(39)	109		
		C(34)-C(39)-C(40)			
		C(34)-C(39)-H(41)			
		C(39)-C(40)-H(42)			
		C(34)-C(39)-H(43)			
		C(39)-C(40)-H(44)			

12.2 QUANTUM-CHEMICAL CALCULATION OF MOLECULE 4-METHOXYINDENE BY METHOD AB INITIO

12.2.1 INTRODUCTION

For the first time it is executed quantum chemical calculation of a molecule of 4-methoxiyndene method AB INITIO with optimization of geometry on all parameters. The optimized geometrical and electronic structure of this compound is received. Acid power of 4-methoxiyndene is theoretically appreciated. It is established, than it to relate to a class of very weak H-acids (pKa = +33, where pKa-universal index of acidity).

12.2.2 AIMS AND BACKGROUNDS

The aim of this work is a study of electronic structure of molecule 4-methoxyindene [1] and theoretical estimation its acid power by quantum-chemical method AB INITIO in base 6–311G**. The calculation was done with optimization of all parameters by standard gradient method built-in in PC GAMESS [2]. The calculation was executed in approach the insulated molecule in gas phase. Program MacMolPlt was used for visual presentation of the model of the molecule [3].

12.2.3 METHODOLOGY

Geometric and electronic structures, general and electronic energies of molecule 4-methoxyindene was received by method AB INITIO in base 6–311G** and are shown on Fig. 12.2 and in Table 12.2. The universal factor of acidity was calculated by formula: pKa = $49.04 - 134.6 * q_{max}^{H+}$ [4, 5] (where, q_{max}^{H+} – a maximum positive charge on atom of the hydrogen q_{max}^{H+} = +0.12 (for 4-methoxyindene q_{max}^{H+} alike Table 12.2)). This same formula is used in Refs. [6–16]. pKa=33.

Quantum-chemical calculation of molecule 4-methoxyindene by method AB INITIO in base 6–311G** was executed for the first time. Optimized geometric and electronic structure of this compound was received. Acid power of molecule 4-methoxyindene was theoretically evaluated (pKa=33). This compound pertains to class of very weak H-acids (pKa>14).

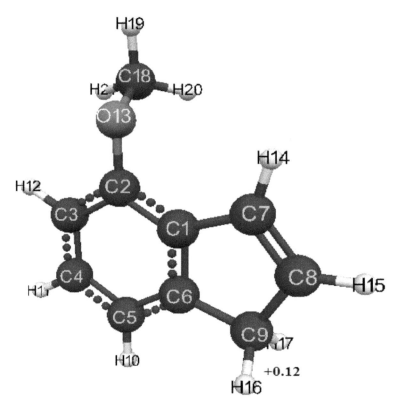

FIGURE. 12.2 Geometric and electronic molecule structure of 4-methoxyindene (E_0= −1,206,038 kDg/mol, E_{el}= −2,721,631 kDg/mol).

TABLE 12.2 Optimized bond lengths, valence corners and charges on atoms of the molecule 4-methoxyindene.

Bond lengths	R, A	Valence corners	Grad	Atom	Charges on atoms
C(2)-C(1)	1.38	C(5)-C(6)-C(1)	121	C(1)	–0.03
C(3)-C(2)	1.39	C(9)-C(6)-C(1)	109	C(2)	+0.22
C(4)-C(3)	1.39	C(1)-C(2)-C(3)	119	C(3)	–0.09
C(5)-C(4)	1.39	O(13)-C(2)-C(3)	121	C(4)	–0.08
C(6)-C(5)	1.38	C(2)-C(3)-C(4)	120	C(5)	–0.08
C(6)-C(1)	1.39	C(3)-C(4)-C(5)	121	C(6)	–0.13
C(6)-C(9)	1.51	C(9)-C(6)-C(5)	130	C(7)	–0.05
C(7)-C(1)	1.47	C(4)-C(5)-C(6)	118	C(8)	–0.15
C(8)-C(7)	1.33	C(2)-C(1)-C(6)	120	C(9)	–0.07
C(9)-C(8)	1.51	C(8)-C(9)-C(6)	102	H(10)	+0.08
H(10)-C(5)	1.08	C(2)-C(1)-C(7)	131	H(11)	+0.09
H(11)-C(4)	1.08	C(1)-C(7)-C(8)	109	H(12)	+0.09
H(12)-C(3)	1.08	C(7)-C(8)-C(9)	111	O(13)	–0.49
O(13)-C(2)	1.36	C(4)-C(5)-H(10)	120	H(14)	+0.09
H(14)-C(7)	1.07	C(3)-C(4)-H(11)	119	H(15)	+0.10
H(15)-C(8)	1.07	C(2)-C(3)-H(12)	119	H(16)	+0.12
H(16)-C(9)	1.09	C(1)-C(2)-O(13)	120	H(17)	+0.12
H(17)-C(9)	1.09	C(1)-C(7)-H(14)	124	C(18)	0.00
C(18)-O(13)	1.41	C(7)-C(8)-H(15)	126	H(19)	+0.10
H(19)-C(18)	1.08	C(8)-C(9)-H(16)	112	H(20)	+0.08
H(20)-C(18)	1.09	C(8)-C(9)-H(17)	112	H(21)	+0.08
H(21)-C(18)	1.09	C(2)-O(13)-C(18)	116		
		O(13)-C(18)-H(19)	107		
		O(13)-C(18)-H(20)	111		
		O(13)-C(18)-H(21)	111		

12.3 QUANTUM-CHEMICAL CALCULATION OF MOLECULE 6-METHOXYINDENE BY METHOD AB INITIO

12.3.1 INTRODUCTION

For the first time it is executed quantum chemical calculation of a molecule of 6-methoxiindene method AB INITIO with optimization of geometry on all parameters. The optimized geometrical and electronic structure of this compound is received. Acid power of 6-methoxiindene is theoretically appreciated. It is established, than it to relate to a class of very weak H-acids (pKa=+32, where pKa-universal index of acidity).

12.3.2 AIMS AND BACKGROUNDS

The aim of this work is a study of electronic structure of molecule 6-methoxyindene [1] and theoretical estimation its acid power by quantum-chemical method AB INITIO in base 6–311G**. The calculation was done with optimization of all parameters by standard gradient method built-in in PC GAMESS [2]. The calculation was executed in approach the insulated molecule in gas phase. Program MacMolPlt was used for visual presentation of the model of the molecule [3].

12.3.3 METHODOLOGY

Geometric and electronic structures, general and electronic energies of molecule 6-methoxyindene was received by method AB INITIO in base 6–311G** and are shown on Fig. 12.3 and in Table 12.3. The universal factor of acidity was calculated by formula: $pKa = 49.04 - 134.6 * q_{max}^{H+}$ [4, 5] (where, q_{max}^{H+} – a maximum positive charge on atom of the hydrogen $q_{max}^{H+}=+0.13$ (for 6-methoxyindene q_{max}^{H+} alike Table 12.3)). This same formula is used in Refs. [6–16] (pKa=32).

Quantum-chemical calculation of molecule 6-methoxyindene by method AB INITIO in base 6–311G** was executed for the first time. Optimized geometric and electronic structure of this compound was received. Acid power of molecule 6-methoxyindene was theoretically evaluated (pKa=32). This compound pertains to class of very weak H-acids (pKa>14).

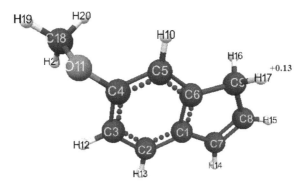

FIGURE 12.3 Geometric and electronic molecule structure of 6-methoxyindene ($E_0 =$ −1,206,038 kDg/mol, $E_{el} = -2,697,224$ kDg/mol).

TABLE 12.3 Optimized bond lengths, valence corners and charges on atoms of the molecule 6-methoxyindene.

Bond lengths	R, A	Valence corners	Grad	Atom	Charges on atoms
C(2)-C(1)	1.38	C(5)-C(6)-C(1)	121	C(1)	−0.05
C(3)-C(2)	1.39	C(1)-C(2)-C(3)	119	C(2)	−0.05
C(4)-C(3)	1.39	C(2)-C(3)-C(4)	120	C(3)	−0.10
C(5)-C(4)	1.39	C(3)-C(4)-C(5)	121	C(4)	+0.22
C(6)-C(5)	1.38	O(11)-C(4)-C(5)	119	C(5)	−0.08
C(6)-C(1)	1.40	C(4)-C(5)-C(6)	119	C(6)	−0.13
C(7)-C(1)	1.47	C(2)-C(1)-C(6)	120	C(7)	−0.06
C(8)-C(7)	1.33	C(2)-C(1)-C(7)	132	C(8)	−0.15
C(8)-C(9)	1.51	C(9)-C(8)-C(7)	111	C(9)	−0.07
C(9)-C(6)	1.51	C(1)-C(7)-C(8)	110	H(10)	+0.09
H(10)-C(5)	1.08	C(6)-C(9)-C(8)	102	O(11)	−0.49
O(11)-C(4)	1.36	C(5)-C(6)-C(9)	130	H(12)	+0.10
H(12)-C(3)	1.08	C(1)-C(6)-C(9)	109	H(13)	+0.09
H(13)-C(2)	1.08	C(4)-C(5)-H(10)	119	H(14)	+0.09
H(14)-C(7)	1.07	C(3)-C(4)-O(11)	120	H(15)	+0.10
H(15)-C(8)	1.07	C(2)-C(3)-H(12)	121	H(16)	+0.12
H(16)-C(9)	1.09	C(1)-C(2)-H(13)	121	**H(17)**	**+0.13**
H(17)-C(9)	1.09	C(1)-C(7)-H(14)	124	C(18)	0.00
C(18)-O(11)	1.41	C(7)-C(8)-H(15)	126	H(19)	+0.10
H(19)-C(18)	1.08	C(9)-C(8)-H(15)	123	H(20)	+0.07
H(20)-C(18)	1.09	C(6)-C(9)-H(16)	112	H(21)	+0.08

TABLE 12.3   *(Continued)*

Bond lengths	R, A	Valence corners	Grad	Atom	Charges on atoms
H(21)-C(18)	1.09	C(6)-C(9)-H(17)	112		
		C(4)-O(11)-C(18)	116		
		O(11)-C(18)-H(19)	107		
		O(11)-C(18)-H(20)	111		
		O(11)-C(18)-H(21)	111		

12.4   QUANTUM-CHEMICAL CALCULATION OF MOLECULE 1-METHYLBICYCLO[4,1,0]HEPTANE BY METHOD AB INITIO.

12.4.1   INTRODUCTION

For the first time quantum chemical calculation of a molecule of 1-methylbicyclo[4,1,0]heptane is executed by method AB INITIO in base 6–311G** with optimization of geometry on all parameters. The optimized geometrical and electronic structure of this compound is received. Acid power of 1-methylbicyclo[4,1,0]heptane is theoretically appreciated. It is established, than it to relate to a class of very weak H-acids (pKa = +34, where pKa-universal index of acidity).

12.4.2   AIMS AND BACKGROUNDS

The aim of this work is a study of electronic structure of molecule 1-methylbicyclo[4,1,0]heptane [1] and theoretical estimation its acid power by quantum-chemical method AB INITIO in base 6–311G**. The calculation was done with optimization of all parameters by standard gradient method built-in in PC GAMESS [2]. The calculation was executed in approach the insulated molecule in gas phase. Program MacMolPlt was used for visual presentation of the model of the molecule [3].

12.4.3 METHODOLOGY

Geometric and electronic structures, general and electronic energies of molecule 1-methylbicyclo[4,1,0]heptane was received by method AB INITIO in base 6–311G** and are shown on Fig. 12.4 and in Table 12.4. The universal factor of acidity was calculated by formula: pKa = $49.04 - 134.6 * q_{max}^{H^+}$ [4, 5] (where, $q_{max}^{H^+}$ – a maximum positive charge on atom of the hydrogen $q_{max}^{H^+} = +0.11$ (for 1-methylbicyclo[4,1,0]heptane $q_{max}^{H^+}$ alike Table 12.4)). This same formula is used in Refs. [6–16]. pKa=34.

Quantum-chemical calculation of molecule 1-methylbicyclo[4,1,0] heptane by method AB INITIO in base 6–311G** was executed for the first time. Optimized geometric and electronic structure of this compound was received. Acid power of molecule 1-methylbicyclo[4,1,0]heptane was theoretically evaluated (pKa=34.). This compound pertains to class of very weak H-acids (pKa>14).

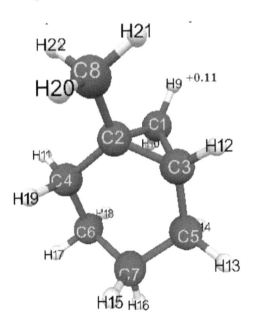

FIGURE 12.4 Geometric and electronic molecule structure of 1-methylbicyclo[4,1,0] heptane ($E_0 = -815,467$ kDg/mol, $E_{el} = -1,862,050$ kDg/mol).

TABLE 12.4 Optimized bond lengths, valence corners and charges on atoms of the molecule 1-methylbicyclo[4,1,0]heptane.

Bond lengths	R, A	Valence corners	Grad	Atom	Charges on atoms
C(2)-C(1)	1.50	C(3)-C(2)-C(1)	60	C(1)	-0.16
C(2)-C(3)	1.51	C(1)-C(3)-C(2)	60	C(2)	−0.23
C(3)-C(1)	1.51	C(5)-C(3)-C(2)	122	C(3)	−0.18
C(4)-C(2)	1.52	C(2)-C(1)-C(3)	60	C(4)	−0.10
C(5)-C(3)	1.53	C(7)-C(5)-C(3)	113	C(5)	−0.14
C(5)-C(7)	1.53	C(1)-C(2)-C(4)	119	C(6)	−0.22
C(6)-C(4)	1.53	C(3)-C(2)-C(4)	118	C(7)	−0.16
C(7)-C(6)	1.53	C(8)-C(2)-C(4)	114	C(8)	−0.16
C(8)-C(2)	1.52	C(1)-C(3)-C(5)	120	**H(9)**	**+0.11**
H(9)-C(1)	1.08	C(6)-C(7)-C(5)	112	H(10)	+0.11
H(10)-C(1)	1.08	C(2)-C(4)-C(6)	114	H(11)	+0.10
H(11)-C(4)	1.09	C(4)-C(6)-C(7)	111	H(12)	+0.11
H(12)-C(3)	1.08	C(1)-C(2)-C(8)	117	H(13)	+0.10
H(13)-C(5)	1.09	C(3)-C(2)-C(8)	118	H(14)	+0.10
H(14)-C(5)	1.09	C(2)-C(1)-H(9)	118	H(15)	+0.09
H(15)-C(7)	1.09	C(2)-C(1)-H(10)	119	H(16)	+0.09
H(16)-C(7)	1.09	C(2)-C(4)-H(11)	109	H(17)	+0.10
H(17)-C(6)	1.09	C(1)-C(3)-H(12)	115	H(18)	+0.09
H(18)-C(6)	1.09	C(3)-C(5)-H(13)	109	H(19)	+0.09
H(19)-C(4)	1.09	C(7)-C(5)-H(13)	109	H(20)	+0.09
H(20)-C(8)	1.09	C(3)-C(5)-H(14)	109	H(21)	+0.08
H(21)-C(8)	1.09	C(6)-C(7)-H(15)	109	H(22)	+0.09
H(22)-C(8)	1.09	C(6)-C(7)-H(16)	111		
		C(4)-C(6)-H(17)	109		
		C(4)-C(6)-H(18)	111		
		C(2)-C(4)-H(19)	108		
		C(2)-C(8)-H(20)	111		
		C(2)-C(8)-H(21)	111		
		C(2)-C(8)-H(22)	111		

12.5   QUANTUM-CHEMICAL CALCULATION OF MOLECULE 1-METHYLBICYCLO[10,1,0]TRIDECANE BY METHOD AB INITIO

12.5.1   INTRODUCTION

For the first time quantum chemical calculation of a molecule of 1-methylbicyclo[10,1,0]tridecane is executed by method AB INITIO in base 6–311G** with optimization of geometry on all parameters. The optimized geometrical and electronic structure of this compound is received. Acid power of 1-methylbicyclo[10,1,0]tridecane is theoretically appreciated. It is established, than it to relate to a class of very weak H-acids (pKa=+33, where pKa-universal index of acidity).

12.5.2   AIMS AND BACKGROUNDS

The aim of this work is a study of electronic structure of molecule 1-methylbicyclo[10,1,0]tridecane [1] and theoretical estimation its acid power by quantum-chemical method AB INITIO in base 6–311G**. The calculation was done with optimization of all parameters by standard gradient method built-in in PC GAMESS [2]. The calculation was executed in approach the insulated molecule in gas phase. Program MacMolPlt was used for visual presentation of the model of the molecule [3].

12.5.3   METHODOLOGY

Geometric and electronic structures, general and electronic energies of molecule 1-methylbicyclo[10,1,0]tridecane was received by method AB INITIO in base 6–311G** and are shown on Fig. 12.5 and in Table 12.5. The universal factor of acidity was calculated by formula: pKa = 49.04–134.6*q_{max}^{H+} [4, 5] (where, q_{max}^{H+} – a maximum positive charge on atom of the hydrogen q_{max}^{H+}=+0.12 (for 1-methylbicyclo[10,1,0]tridecane q_{max}^{H+} alike Table 12.5)). This same formula is used in Refs. [6–16]. pKa=33.

Quantum-chemical calculation of molecule 1-methy-lbicyclo[10,1,0] tridecane by method AB INITIO in base 6–311G** was executed for the first time. Optimized geometric and electronic structure of this compound was received. Acid power of molecule 1-methylbicyclo[10,1,0]tridecane was theoretically evaluated (pKa=33.). This compound pertains to class of very weak H-acids (pKa>14).

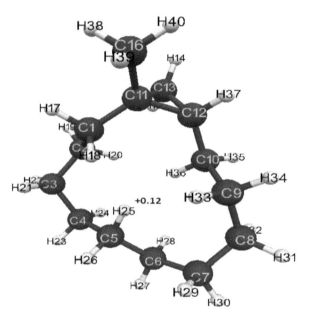

FIGURE 12.5 Geometric and electronic molecule structure of 1-methylbicyclo[10,1,0] tridecane ($E_0 = -1,429,365$ kDg/mol, $E_{el} = -4,000,827$ kDg/mol).

TABLE 12.5 Optimized bond lengths, valence corners and charges on atoms of the molecule 1-methylbicyclo[10,1,0]tridecane.

Bond lengths	R,A	Valence corners	Grad	Atom	Charges on atoms
C(2)-C(1)	1.54	C(1)-C(2)-C(3)	115	C(1)	-0.13
C(3)-C(2)	1.55	C(2)-C(3)-C(4)	113	C(2)	-0.21
C(4)-C(3)	1.54	C(3)-C(4)-C(5)	114	C(3)	-0.17
C(5)-C(4)	1.53	C(4)-C(5)-C(6)	115	C(4)	-0.18
C(6)-C(5)	1.53	C(5)-C(6)-C(7)	114	C(5)	-0.20

TABLE 12.5 *(Continued)*

Bond lengths	R, A	Valence corners	Grad	Atom	Charges on atoms
C(8)-C(7)	1.54	C(7)-C(8)-C(9)	119	C(7)	−0.18
C(9)-C(8)	1.55	C(8)-C(9)-C(10)	115	C(8)	−0.17
C(10)-C(9)	1.54	C(11)-C(12)-C(10)	126	C(9)	−0.20
C(10)-C(12)	1.52	C(2)-C(1)-C(11)	119	C(10)	−0.15
C(11)-C(1)	1.54	C(1)-C(11)-C(12)	123	C(11)	−0.21
C(12)-C(11)	1.51	C(11)-C(13)-C(12)	240	C(12)	−0.14
C(12)-C(13)	1.51	C(1)-C(11)-C(13)	123	C(13)	−0.16
C(13)-C(11)	1.50	C(11)-C(13)-H(14)	118	H(14)	+0.11
H(14)-C(13)	1.08	C(11)-C(13)-H(15)	119	H(15)	+0.11
H(15)-C(13)	1.08	C(1)-C(11)-C(16)	112	C(16)	−0.17
C(16)-C(11)	1.52	C(2)-C(1)-H(17)	107	H(17)	+0.10
H(17)-C(1)	1.09	C(2)-C(1)-H(18)	111	H(18)	+0.10
H(18)-C(1)	1.09	C(1)-C(2)-H(19)	108	H(19)	+0.10
H(19)-C(2)	1.09	C(1)-C(2)-H(20)	111	H(20)	+0.09
H(20)-C(2)	1.08	C(2)-C(3)-H(21)	110	H(21)	+0.09
H(21)-C(3)	1.09	C(2)-C(3)-H(22)	109	H(22)	+0.10
H(22)-C(3)	1.09	C(3)-C(4)-H(23)	110	H(23)	+0.09
H(23)-C(4)	1.09	C(3)-C(4)-H(24)	108	H(24)	+0.09
H(24)-C(4)	1.09	C(4)-C(5)-H(25)	109	**H(25)**	**+0.12**
H(25)-C(5)	1.08	C(4)-C(5)-H(26)	109	H(26)	+0.09
H(26)-C(5)	1.09	C(5)-C(6)-H(27)	109	H(27)	+0.09
H(27)-C(6)	1.09	C(5)-C(6)-H(28)	110	H(28)	+0.08
H(28)-C(6)	1.09	C(6)-C(7)-H(29)	109	H(29)	+0.09
H(29)-C(7)	1.09	C(6)-C(7)-H(30)	108	H(30)	+0.09
H(30)-C(7)	1.09	C(7)-C(8)-H(31)	106	H(31)	+0.09
H(31)-C(8)	1.09	C(7)-C(8)-H(32)	109	H(32)	+0.09
H(32)-C(8)	1.09	C(8)-C(9)-H(33)	110	H(33)	+0.09
H(33)-C(9)	1.09	C(8)-C(9)-H(34)	108	H(34)	+0.09
H(34)-C(9)	1.09	C(9)-C(10)-H(35)	108	H(35)	+0.10
H(35)-C(10)	1.09	C(9)-C(10)-H(36)	111	H(36)	+0.09
H(36)-C(10)	1.08	C(11)-C(12)-H(37)	113	H(37)	+0.10
H(37)-C(12)	1.08	C(11)-C(16)-H(38)	111	H(38)	+0.09
H(38)-C(16)	1.09	C(11)-C(16)-H(39)	111	H(39)	+0.09
H(39)-C(16)	1.09	C(11)-C(16)-H(40)	112	H(40)	+0.08
H(40)-C(16)	1.09				

12.6   QUANTUM-CHEMICAL CALCULATION OF MOLECULE 2,11-SPIROTETRADECANE BY METHOD AB INITIO

12.6.1   INTRODUCTION

For the first time quantum chemical calculation of a molecule of 2,11-spirotetradecane is executed by method AB INITIO in base 6–311G** with optimization of geometry on all parameters. The optimized geometrical and electronic structure of this compound is received. Acid power of 2,11-spirotetradecane. It is established, than it to relate to a class of very weak H-acids (pKa = +34, where pKa-universal index of acidity).

12.6.2   AIMS AND BACKGROUNDS

The aim of this work is a study of electronic structure of molecule 2,11-spirotetradecane [1] and theoretical estimation its acid power by quantum-chemical method AB INITIO in base 6–311G**. The calculation was done with optimization of all parameters by standard gradient method built-in in PC GAMESS [2]. The calculation was executed in approach the insulated molecule in gas phase. Program MacMolPlt was used for visual presentation of the model of the molecule [3].

12.6.3   METHODOLOGY

Geometric and electronic structures, general and electronic energies of molecule 2,11-spirotetradecane was received by method AB INITIO in base 6–311G** and are shown on Fig. 12.6 and in Table 12.6. The universal factor of acidity was calculated by formula: pKa = $49.04 - 134.6 * q_{max}^{H+}$ [4, 5] (where, q_{max}^{H+} – a maximum positive charge on atom of the hydrogen $q_{max}^{H+} = +0.11$ (for 2,11-spirotetradecane q_{max}^{H+} alike Table 12.6)). This same formula is used in Refs. [6–16]. pKa=34.

Quantum-chemical calculation of molecule 2,11-spirotetradecane by method AB INITIO in base 6–311G** was executed for the first time. Optimized geometric and electronic structure of this compound was received. Acid power of molecule 2,11-spirotetradecane was theoretically evaluated (pKa=34.). This compound pertains to class of very weak H-acids (pKa>14).

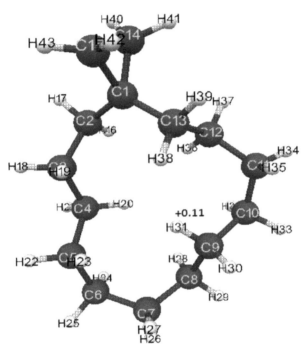

FIGURE 12.6 Geometric and electronic molecule structure of 2,11-spirotetradecane ($E_0 = -1,531,699$ kDg/mol, $E_{эл} = -4,362,610$ kDg/mol).

TABLE 12.6 Optimized bond lengths, valence corners and charges on atoms of the molecule 2,11-spirotetradecane.

Bond lengths	R, A	Valence corners	Grad	Atom	Charges on atoms
C(2)-C(1)	1.53	C(1)-C(2)-C(3)	114	C(1)	-0.24
C(3)-C(2)	1.55	C(2)-C(3)-C(4)	114	C(2)	-0.12
C(4)-C(3)	1.54	C(3)-C(4)-C(5)	113	C(3)	-0.18
C(5)-C(4)	1.53	C(4)-C(5)-C(6)	116	C(4)	-0.18
C(6)-C(5)	1.54	C(5)-C(6)-C(7)	118	C(5)	-0.19
C(7)-C(6)	1.54	C(6)-C(7)-C(8)	118	C(6)	-0.17

TABLE 12.6   *(Continued)*

Bond lengths	R, A	Valence corners	Grad	Atom	Charges on atoms
C(8)-C(7)	1.54	C(7)-C(8)-C(9)	116	C(6)	−0.18
C(9)-C(8)	1.53	C(8)-C(9)-C(10)	113	C(7)	−0.18
C(10)-C(9)	1.53	C(9)-C(10)-C(11)	115	C(8)	−0.19
C(11)-C(10)	1.54	C(10)-C(11)-C(12)	114	C(9)	−0.18
C(12)-C(11)	1.55	C(1)-C(13)-C(12)	115	C(10)	−0.18
C(12)-C(13)	1.53	C(2)-C(1)-C(13)	117	C(11)	−0.17
C(13)-C(1)	1.53	C(2)-C(1)-C(14)	117	C(12)	−0.12
C(14)-C(1)	1.50	C(1)-C(15)-C(14)	240	C(13)	−0.18
C(14)-C(15)	1.50	C(2)-C(1)-C(15)	116	C(14)	−0.18
C(15)-C(1)	1.50	C(1)-C(2)-H(16)	110	C(15)	+0.09
H(16)-C(2)	1.09	C(1)-C(2)-H(17)	108	H(16)	+0.08
H(17)-C(2)	1.09	C(2)-C(3)-H(18)	110	H(17)	+0.09
H(18)-C(3)	1.09	C(2)-C(3)-H(19)	109	H(18)	+0.09
H(19)-C(3)	1.09	C(3)-C(4)-H(20)	110	H(19)	+0.09
H(20)-C(4)	1.08	C(3)-C(4)-H(21)	109	H(20)	+0.09
H(21)-C(4)	1.09	C(4)-C(5)-H(22)	108	H(21)	+0.09
H(22)-C(5)	1.09	C(4)-C(5)-H(23)	109	H(22)	+0.09
H(23)-C(5)	1.09	C(5)-C(6)-H(24)	109	H(23)	+0.09
H(24)-C(6)	1.09	C(5)-C(6)-H(25)	108	H(24)	+0.09
H(25)-C(6)	1.09	C(6)-C(7)-H(26)	106	H(25)	+0.09
H(26)-C(7)	1.09	C(6)-C(7)-H(27)	109	H(26)	+0.09
H(27)-C(7)	1.09	C(7)-C(8)-H(28)	109	H(27)	+0.09
H(28)-C(8)	1.09	C(7)-C(8)-H(29)	109	H(28)	+0.09
H(29)-C(8)	1.09	C(8)-C(9)-H(30)	109	H(29)	+0.09
H(30)-C(9)	1.09	C(8)-C(9)-H(31)	109	H(30)	**+0.11**
H(31)-C(9)	1.08	C(9)-C(10)-H(32)	109	**H(31)**	+0.09
H(32)-C(10)	1.09	C(9)-C(10)-H(33)	109	H(32)	+0.09
H(33)-C(10)	1.09	C(10)-C(11)-H(34)	108	H(33)	+0.10
H(34)-C(11)	1.09	C(10)-C(11)-H(35)	110	H(34)	+0.09
H(35)-C(11)	1.09	C(11)-C(12)-H(36)	109	H(35)	+0.09
H(36)-C(12)	1.09	C(11)-C(12)-H(37)	109	H(36)	+0.10
H(37)-C(12)	1.09	C(1)-C(13)-H(38)	109	H(37)	+0.10
H(38)-C(13)	1.09	C(1)-C(13)-H(39)	108	H(38)	+0.09
H(39)-C(13)	1.09	C(1)-C(14)-H(40)	119	H(39)	+0.11
H(40)-C(14)	1.08	C(1)-C(14)-H(41)	118	H(40)	+0.11
H(41)-C(14)	1.08	C(1)-C(15)-H(42)	119	H(41)	+0.11
H(42)-C(15)	1.08	C(1)-C(15)-H(43)	118	H(42)	+0.11
H(43)-C(15)	1.08			H(43)	

12.7 QUANTUM-CHEMICAL CALCULATION OF MOLECULE 6,6-DIMETHYLFULVENE BY METHOD MNDO

12.7.1 INTRODUCTION

For the first time quantum chemical calculation of a molecule of 6,6-di-methylfulvene is executed by method MNDO with optimization of geometry on all parameters. The optimized geometrical and electronic structure of this compound is received. Acid power of 6,6-dimethylfulvene is theoretically appreciated. It is established, than it to relate to a class of very weak H-acids (pKa=+30, where pKa-universal index of acidity).

12.7.2. AIMS AND BACKGROUNDS

The aim of this work is a study of electronic structure of molecule 6,6-dimethylfulvene [1] and theoretical estimation its acid power by quantum-chemical method MNDO. The calculation was done with optimization of all parameters by standard gradient method built-in in PC GAMESS [2]. The calculation was executed in approach the insulated molecule in gas phase. Program MacMol-Plt was used for visual presentation of the model of the molecule [3].

12.7.3 METHODOLOGY

Geometric and electronic structures, general and electronic energies of molecule 6,6-dimethylfulvene was received by method MNDO and are shown on Fig. 12.7 and in Table 12.7. The universal factor of acidity was calculated by formula: $pKa = 42.11 - 147.18 * q_{max}^{H+}$ [4, 5] (where, q_{max}^{H+} – a maximum positive charge on atom of the hydrogen $q_{max}^{H+} = +0.07$ (for 6,6-dimethylfulvene q_{max}^{H+} alike Table 12.7)). This same formula is used in Refs. [17–27]. pKa=33.

Quantum-chemical calculation of molecule 6,6-dimethylfulvene by method MNDO was executed for the first time. Optimized geometric and electronic structure of this compound was received. Acid power of molecule 6,6-dimethylfulvene was theoretically evaluated (pKa=33). This compound pertains to class of very weak H-acids (pKa>14).

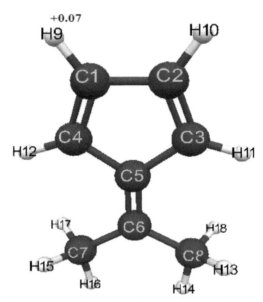

FIGURE 12.7   Geometric and electronic molecule structure of 6,6-dimethylfulvene ($E_0=$ −515,518.5 kDg/mol, $E_{el}=$ −112,208.25 kDg/mol).

TABLE 12.7   Optimized bond lengths, valence corners and charges on atoms of the molecule 6,6–6,6-dimethylfulvene.

Bond lengths	R, A	Valence corners	Grad	Atom	Charges on atoms
C(2)-C(1)	1.47	C(4)-C(1)-C(2)	109	C(1)	−0.07
C(3)-C(2)	1.37	C(1)-C(2)-C(3)	109	C(2)	−0.07
C(4)-C{1)	1.36	C(1)-C(4)-C(5)	109	C(3)	−0.08
C(5)-C(3)	1.49	C(2)-C(3)-C(5)	109	C(4)	−0.08
C(5)-C(4)	1.49	C(3)-C(5)-C(4)	104	C(5)	−0.06
C(6)-C(5)	1.36	C(3)-C(5)-C(6)	128	C(6)	−0.06
C(6)-C(7)	1.51	C(4)-C(5)-C(6)	128	C(7)	+0.07
C(6)-C(8)	1.51	C(5)-C(6)-C(7)	122	C(8)	+0.07
H(9)-C(1)	1.08	C(5)-C(6)-C(8)	122	**H(9)**	**+0.07**
H(10)-C(2)	1.08	C(4)-C(1)-H(9)	128	H(10)	+0.07
H(11)-C(3)	1.08	C(1)-C(2)-H(10)	124	H(11)	+0.07
H(12)-C(4)	1.08	C(2)-C(3)-H(11)	126	H(12)	+0.07
H(13)-C(8)	1.11	C(5)-C(4)-H(12)	124	H(13)	0.00
H(14)-C(8)	1.11	C(6)-C(8)-H(13)	111	H (14)	0.00

TABLE 12.7   *(Continued)*

Bond lengths	R, A	Valence corners	Grad	Atom	Charges on atoms
H(15)-C(7)	1.11	C(6)-C(8)-H(14)	113	H (15)	0.00
H(16)-C(7)	1.11	C(6)-C(7)-H(15)	111	H(16)	0.00
H(17)-C(7)	1.11	C(6)-C(7)-H(16)	113	H (17)	0.00
H(18)-C(8)	1.11	C(6)-C(7)-H(17)	111	H(18)	0.00
		C(6)-C(8)-H(18)			

12.8   QUANTUM-CHEMICAL CALCULATION OF MOLECULE CYCLOHEXADIENE-1,3 BY METHOD MNDO
12.8.1. INTRODUCTION

For the first time quantum chemical calculation of a molecule of cyclo-hexadiene-1,3 is executed by method MNDO with optimization of geom-etry on all parameters. The optimized geometrical and electronic structure of this compound is received. Acid power of cyclohexadiene-1,3 is theo-retically appreciated. It is established, than it to relate to a class of very weak H-acids (pKa=+33, where pKa-universal index of acidity).

12.8.2   AIMS AND BACKGROUNDS

The aim of this work is a study of electronic structure of molecule cyclohexa-diene-1,3 [1] and theoretical estimation its acid power by quantum-chemical method MNDO. The calculation was done with optimization of all parameters by standard gradient method built-in in PC GAMESS [2]. The calculation was executed in approach the insulated molecule in gas phase. Program MacMol-Plt was used for visual presentation of the model of the molecule [3].

12.8.3   METHODOLOGY

Geometric and electronic structures, general and electronic energies of molecule cyclohexadiene-1,3 was received by method MNDO and are shown on Fig. 12.8 and in Table 12.8. The universal factor of acidity was calculated by formula: $pKa = 42.11 - 147.18 \cdot q_{max}^{H+}$ [4, 5] (where, q_{max}^{H+} – a maximum positive charge on atom of the hydrogen $q_{max}^{H+} = +0.06$ (for cyclohexadiene-1,3 q_{max}^{H+} alike Table 12.8)). This same formula is used in Refs. [17–27]. pKa=33.

Quantum-chemical calculation of molecule cyclohexadiene-1,3 by method MNDO was executed for the first time. Optimized geometric and electronic structure of this compound was received. Acid power of molecule cyclohexadiene-1,3 was theoretically evaluated (pKa=33). This compound pertains to class of very weak H-acids (pKa>14).

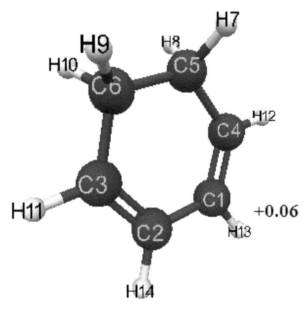

FIGURE 12.8   Geometric and electronic molecule structure of cyclohexadi-ene-1,3 ($E_0 = -84,906$ kDg/mol, $E_{el} = -344,668$ kDg/mol)

TABLE 12.8 Optimized bond lengths, valence corners and charges on atoms of the molecule cyclohexadiene-1,3.

Bond lengths	R, A	Valence corners	Grad	Atom	Charges on atoms
C(2)-C(1)	1.46	C(1)-C(2)-C(3)	121	C(1)	−0.06
C(3)-C(2)	1.35	C(2)-C(1)-C(4)	121	C(2)	−0.06
C(4)-C(1)	1.35	C(6)-C(5)-C(4)	116	C(3)	−0.09
C(5)-C(4)	1.50	C(1)-C(4)-C(5)	123	C(4)	−0.09
C(5)-C(6)	1.55	C(3)-C(6)-C(5)	116	C(5)	+0.02
C(6)-C(3)	1.50	C(2)-C(3)-C(6)	123	C(6)	+0.02
H(7)-C(5)	1.12	C(4)-C(5)-H(7)	108	H(7)	+0.01
H(8)-C(5)	1.12	C(6)-C(5)-H(7)	109	H(8)	+0.01
H(9)-C(6)	1.12	C(4)-C(5)-H(8)	108	H(9)	+0.01
H(10)-C(6)	1.12	C(6)-C(5)-H(8)	109	H(10)	+0.01
H(11)-C(3)	1.09	C(3)-C(6)-H(9)	108	H(11)	+0.05
H(12)-C(4)	1.09	C(3)-C(6)-H(10)	108	H(12)	+0.05
H(13)-C(1)	1.09	C(2)-C(3)-H(11)	121	**H(13)**	**+0.06**
H(14)-C(2)	1.09	C(1)-C(4)-H(12)	121	H(14)	+0.06
		C(2)-C(1)-H(13)	117		
		C(1)-C(2)-H(14)	117		

12.9 QUANTUM-CHEMICAL CALCULATION OF MOLECULE ALLYLMETHYLCYCLOPENTADIENE BY METHOD MNDO

12.9.1 INTRODUCTION

For the first time quantum chemical calculation of a molecule of allyl-methylcyclopentadiene is executed by method MNDO with optimization of geometry on all parameters. The optimized geometrical and electronic structure of this compound is received. Acid power of allylmethylcyclo-pentadiene is theoretically appreciated. It is established, than it to relate to a class of very weak H-acids (pKa=+32, where pKa-universal index of acidity).

12.9.2. AIMS AND BACKGROUNDS

The aim of this work is a study of electronic structure of molecule allyl-methylcyclopentadiene [1] and theoretical estimation its acid power by quantum-chemical method MNDO. The calculation was done with optimization of all parameters by standard gradient method built-in in PC GAMESS [2]. The calculation was executed in approach the insulated molecule in gas phase. Program MacMolPlt was used for visual presentation of the model of the molecule [3].

12.9.3 METHODOLOGY

Geometric and electronic structures, general and electronic energies of molecule allylmethylcyclopentadiene were received by method MNDO and are shown on Fig. 12.9 and in Table 12.9. The universal factor of acidity was calculated by formula: $pKa = 42.11 - 147.18 * q_{max}^{H+}$ [4, 5] (where, q_{max}^{H+} – a maximum positive charge on atom of the hydrogen $q_{max}^{H+} = +0.07$ (for allylmethylcyclopentadiene q_{max}^{H+} alike Table 12.9)). This same formula is used in Refs. [17–27]. pKa=32.

Quantum-chemical calculation of molecule allylmethylcyclopentadiene by method MNDO was executed for the first time. Optimized geometric and electronic structure of this compound was received. Acid power of molecule allylmethylcyclopentadiene was theoretically evaluated (pKa=32). This compound pertains to class of very weak H-acids (pKa>14).

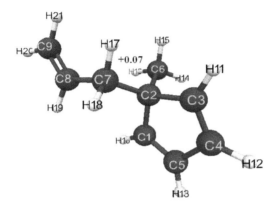

FIGURE 12.9 Geometric and electronic molecule structure of allylmethylcyclopentadiene ($E_0 = -127,244$ kDg/mol, $E_{el} = -640,709$ kDg/mol).

TABLE 12.9 Optimized bond lengths, valence corners and charges on atoms of the molecule allylmethylcyclopentadiene.

Bond lengths	R, A	Valence corners	Grad	Atom	Charges on atoms
C(1)-C(2)	1.54	C(2)-C(1)-H(10)	122	C(1)	−0.09
C(2)-C(3)	1.54	C(1)-C(5)-H(10)	127	C(2)	−0.04
C(3)-C(4)	1.36	C(1)-C(5)-H(13)	128	C(3)	−0.10
C(4)-C(5)	1.47	C(2)-C(3)-H(11)	122	C(4)	−0.07
C(5)C(1)	1.36	C(3)-C(4)-H(12)	128	C(5)	−0.07
C(2)-C(7)	1.56	C(4)-C(5)-H(13)	123	C(6)	0.07
C(2)-C(6)	1.55	C(5)-C(4)-H(12)	123	C(7)	0.07
C(7)-C(8)	1.51	C(4)-C(3)-H(11)	127	C(8)	−0.12
C(8)-C(9)	1.34	C(2)-C(7)-H(17)	108	C(9)	−0.05
H(10)-C(1)	1.08	C(2)-C(7)-H(18)	109	**H(10)**	**0.07**
H(11)-C(3)	1.08	C(7)-C(8)-H(19)	114	H(11)	0.07
H(12)-C(4)	1.08	C(9)-C(8)-H(19)	119	H(12)	0.07
H(13)-C(5)	1.08	C(8)-C(9)-H(20)	122	H(13)	0.07
H(14)-C(6)	1.11	C(8)-C(9)-H(21)	124	H(14)	−0.00
H(15)-C(6)	1.11	C(8)-C(7)-H(17)	110	H(15)	−0.01
H(16)-C(6)	1.11	C(8)-C(7)-H(18)	107	H(16)	−0.00
H(17)-C(7)	1.12	C(2)-C(6)-H(14)	111	H(17)	0.00
H(18)-C(7)	1.12	C(2)-C(6)-H(15)	111	H(18)	0.01
H(19)-C(8)	1.10	C(2)-C(6)-H(16)	112	H(19)	0.05
H(20)-C(9)	1.09			H(20)	0.04
H(21)-C(9)	1.09			H(21)	0.04

12.10   QUANTUM-CHEMICAL CALCULATION OF MOLECULE 1,2-DIMETHYLENCYCLOHEXANE BY METHOD MNDO

12.10.1   INTRODUCTION

For the first time quantum chemical calculation of a molecule of 1,2-di-methylencyclohexane is executed by method MNDO with optimization of geometry on all parameters. The optimized geometrical and electronic structure of this compound is received. Acid power of 1,2-dimethylency-clohexane is theoretically appreciated. It is established, than it to relate to a class of very weak H-acids (pKa=+36, where pKa-universal index of acidity).

12.10.2   AIMS AND BACKGROUNDS

The aim of this work is a study of electronic structure of molecule 1,2-di-methylencyclohexane [1] and theoretical estimation its acid power by quantum-chemical method MNDO. The calculation was done with op-timization of all parameters by standard gradient method built-in in PC GAMESS [2]. The calculation was executed in approach the insulated molecule in gas phase. Program MacMolPlt was used for visual presenta-tion of the model of the molecule [3].

12.110.3   METHODOLOGY

Geometric and electronic structures, general and electronic energies of molecule 1,2-dimethylencyclohexane were received by method MNDO and are shown on Fig. 12.10 and in Table 12.10. The universal factor of acidity was calculated by formula: $pKa = 42.11 - 147.18 * q_{max}^{H+}$ [4, 5] (where, q_{max}^{H+} − a maximum positive charge on atom of the hydrogen $q_{max}^{H+} = +0.04$ (for 1,2-dimethylencyclohexane q_{max}^{H+} alike Table 12.10)). This same formula is used in Refs. [17–27]. pKa=36.

Quantum-chemical calculation of molecule 1,2-dimethylencyclohex-ane by method MNDO was executed for the first time. Optimized geo-metric and electronic structure of this compound was received. Acid power of molecule 11,2-dimethylencyclohexane was theoretically evalu-ated (pKa=36). This compound pertains to class of very weak H-acids (pKa>14).

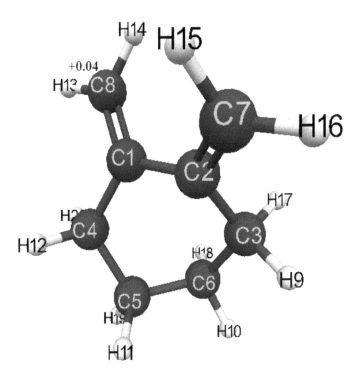

FIGURE12.10   Geometricandelectronicmoleculestructureof1,2-dimethylencyclohexane (E_0= −115,061 kDg/mol, E_{el}= −565,198 kDg/mol).

TABLE 12.10 Optimized bond lengths, valence corners and charges on atoms of the molecule 1,2-dimethylencyclohexane.

Bond lengths	R, A	Valence corners	Grad	Atom	Charges on atoms
(2)-C(1)	1.49	C(1)-C(2)-C(3)	116	C(1)	−0.09
C(3)-C(2)	1.52	C(5)-C(6)-C(3)	114	C(2)	−0.09
C(3)-C(6)	1.54	C(2)-C(1)-C(4)	116	C(3)	+0.04
C(4)-C(1)	1.52	C(1)-C(4)-C(5)	115	C(4)	+0.04
C(5)-C(4)	1.54	C(4)-C(5)-C(6)	114	C(5)	−0.01
C(6)-C(5)	1.54	C(1)-C(2)-C(7)	123	C(6)	−0.01
C(7)-C(2)	1.35	C(2)-C(1)-C(8)	123	C(7)	−0.04
C(8)-C(1)	1.35	C(2)-C(3)-H(9)	110	C(8)	−0.04
H(9)-C(3)	1.11	C(5)-C(6)-H(10)	109	H(9)	0.00
H(10)-C(6)	1.11	C(4)-C(5)-H(11)	109	H(10)	0.00
H(11)-C(5)	1.11	C(1)-C(4)-H(12)	109	H(11)	+0.01
H(12)-C(4)	1.12	C(1)-C(8)-H(13)	123	H(12)	+0.01
H(13)-C(8)	1.09	C(1)-C(8)-H(14)	124	**H(13)**	**+0.04**
H(14)-C(8)	1.09	C(2)-C(7)-H(15)	124	H(14)	+0.04
H(15)-C(7)	1.09	C(2)-C(7)-H(16)	123	H(15)	+0.04
H(16)-C(7)	1.09	C(2)-C(3)-H(17)	109	H(16)	+0.04
H(17)-C(3)	1.12	C(5)-C(6)-H(18)	109	H(17)	+0.01
H(18)-C(6)	1.11	C(4)-C(5)-H(19)	109	H(18)	+0.01
H(19)-C(5)	1.11	C(1)-C(4)-H(20)	110	H(19)	0.00
H(20)-C(4)	1.11			H(20)	0.00

KEYWORDS

- 1-methyl-bicyclo[10,1,0]tridecane
- 1-methylbicyclo[4,1,0]heptane
- 1,2-dimethylencyclohexane
- 2,11-spirotetradecane
- 6,6-dimethylfulvene
- acid power
- allylmethylcyclopentadiene
- cyclohexadiene-1,3
- method AB INITIO
- method MNDO
- quantum chemical calculation

REFERENCES

1. Kennedi, J. Cationic polymerization of olefins. Moscow, 1978. 431 p.
2. Shmidt, M. W., Baldrosge, K. K., Elbert, J. A., Gordon, M. S., Enseh, J. H., Koseki, S., N.Matsvnaga., Nguyen, K. A., SU, S. J., and anothers. J. Comput. Chem.14, 1347–1363, (1993).
3. Bode, B. M., Gordon, M. S., J. Mol. Graphics Mod., 16, 1998, 133–138.
4. Babkin, V. A., Fedunov, R. G., K. S. Minsker and anothers. Oxidation communication, 2002, №1, 25, 21–47.
5. Babkin, V. A. et al. Oxidation communication, 21, №4, 1998, 454–460.
6. Babkin, V. A., Dmitriev, V. Yu., G. E. Zaikov. Geometrical and electronic structure of molecule benzilpenicillin by method AB INITIO. In book: Quantum-chemical calculations of molecular system as the basis of nanotechnologies in applied quantum chemistry. Volume, I. New York, Nova Publisher, 2012, 7–10.
7. Babkin, V. A., A. B. Tsykanov. Geometrical and electronic structure of molecule cellulose by method AB INITIO. In book: Quantum-chemical calculations of molecular system as the basis of nanotechnologies in applied quantum chemistry. Volume, I. New York, Nova Publisher, 2012, 31–34.
8. Babkin, V. A., Dmitriev, V. Yu., G. E. Zaikov. Geometrical and electronic structure of molecule aniline by method AB INITIO. In book: Quantum-chemical calculations of molecular system as the basis of nanotechnologies in applied quantum chemistry. Volume, I. New York, Nova Publisher, 2012, 89–91.

9. Babkin, V. A., Dmitriev, V. Yu., G. E. Zaikov. Geometrical and electronic structure of molecule butene-1 by method AB INITIO. In book: Quantum-chemical calculations of molecular system as the basis of nanotechnologies in applied quantum chemistry. Volume, I. New York, Nova Publisher, 2012, 109–111.

10. Babkin, V. A., Dmitriev, V. Yu., G. E. Zaikov. Geometrical and electronic structure of molecule butene-2 by method AB INITIO. In book: Quantum-chemical calculations of molecular system as the basis of nanotechnologies in applied quantum chemistry. Volume, I. New York, Nova Publisher, 2012, 113–115.

11. Babkin, V. A., V. V. Galenkin. Geometrical and electronic structure of molecule 3, 3-dimethylbutene-1 by method AB INITIO. In book: Quantum-chemical calculations of molecular system as the basis of nanotechnologies in applied quantum chemistry. Volume, I. New York, Nova Publisher, 2012, 129–131.

12. Babkin, V. A., D. S. Andreev. Geometrical and electronic structure of molecule 4, 4-dimethylpentene-1 by method AB INITIO. In book: Quantum-chemical calculations of molecular system as the basis of nanotechnologies in applied quantum chemistry. Volume, I. New York, Nova Publisher, 2012, 141–143.

13. Babkin, V. A., D. S. Andreev. Geometrical and electronic structure of molecule 4-methylhexene-1 by method AB INITIO. In book: Quantum-chemical calculations of molecular system as the basis of nanotechnologies in applied quantum chemistry. Volume, I. New York, Nova Publisher, 2012, 145–147.

14. Babkin, V. A., D. S. Andreev. Geometrical and electronic structure of molecule 4-methylpentene-1 by method AB INITIO. In book: Quantum-chemical calculations of molecular system as the basis of nanotechnologies in applied quantum chemistry. Volume, I. New York, Nova Publisher, 2012, 149–151.

15. Babkin, V. A., D. S. Andreev. Geometrical and electronic structure of molecule isobutylene by method AB INITIO. In book: Quantum-chemical calculations of molecular system as the basis of nanotechnologies in applied quantum chemistry. Volume, I. New York, Nova Publisher, 2012, 155–157.

16. Babkin, V. A., D. S. Andreev. Geometrical and electronic structure of molecule 2-methylbutene-1 by method AB INITIO. In book: Quantum-chemical calculations of molecular system as the basis of nanotechnologies in applied quantum chemistry. Volume, I. New York, Nova Publisher, 2012, 159–161.

17. Babkin, V. A., Andreev, D. S., E. S. Titova. Geometrical and electronic structure of molecule vitamin "C" by method MNDO. In book: Quantum-chemical calculations of molecular system as the basis of nanotechnologies in applied quantum chemistry. Volume, I. New York, Nova Publisher, 2012, 3–5

18. Babkin, V. A., Andreev, D. S., Titova, E. S., G. E. Zaikov. Geometrical and electronic structure of molecule vitamin "A" by method MNDO. In book: Quantum-chemical calculations of molecular system as the basis of nanotechnologies in applied quantum chemistry. Volume, I. New York, Nova Publisher, 2012, 11–14.

19. Babkin, V. A., D. S. Andreev. Geometrical and electronic structure of molecules nematic, smectic, holesteric liquid crystal by method MNDO. In book: Quantum-chemical calculations of molecular system as the basis of nanotechnologies in applied quantum chemistry. Volume, I. New York, Nova Publisher, 2012, 17–24.

20. Babkin, V. A., Dmitriev, V. Yu., Andreev, D. S., G. E. Zaikov. Geometrical and electronic structure of molecule holesterinbenzoat by method MNDO. In book: Quantum-

chemical calculations of molecular system as the basis of nanotechnologies in applied quantum chemistry. Volume, I. New York, Nova Publisher, 2012, 25–28.

21. Babkin, V. A., Dmitriev, V. Yu., G. E. Zaikov. Geometrical and electronic structure of molecule butene-1 by method MNDO. In book: Quantum-chemical calculations of molecular system as the basis of nanotechnologies in applied quantum chemistry. Volume, I. New York, Nova Publisher, 2012, 119–121.

22. Babkin, V. A., Dmitriev, V. Yu., G. E. Zaikov. Geometrical and electronic structure of molecule butene-2 by method MNDO. In book: Quantum-chemical calculations of molecular system as the basis of nanotechnologies in applied quantum chemistry. Volume, I. New York, Nova Publisher, 2012, 123–125.

23. Babkin, V. A., V. V. Galenkin. Geometrical and electronic structure of molecule 3, 3-dimethylbutene-1 by method MNDO. In book: Quantum-chemical calculations of molecular system as the basis of nanotechnologies in applied quantum chemistry. Volume, I. New York, Nova Publisher, 2012, 135–137.

24. Babkin, V. A., D. S. Andreev. Geometrical and electronic structure of molecule iso-butylene by method MNDO. In book: Quantum-chemical calculations of molecular system as the basis of nanotechnologies in applied quantum chemistry. Volume, I. New York, Nova Publisher, 2012, 165–167.

25. Babkin, V. A., D. S. Andreev. Geometrical and electronic structure of molecule 2-meth-ylbutene-1 by method MNDO. In book: Quantum-chemical calculations of molecular system as the basis of nanotechnologies in applied quantum chemistry. Volume, I. New York, Nova Publisher, 2012, 169–171.

26. Babkin, V. A., D. S. Andreev. Geometrical and electronic structure of molecule buta-diene-1, 3 by method MNDO. In book: Quantum-chemical calculations of molecular system as the basis of nanotechnologies in applied quantum chemistry. Volume, I. New York, Nova Publisher, 2012, 193–195.

27. Babkin, V. A., D. S. Andreev. Geometrical and electronic structure of molecule 2-meth-ylbutadiene-1, 3 by method MNDO. In book: Quantum-chemical calculations of mo-lecular system as the basis of nanotechnologies in applied quantum chemistry. Volume, I. New York, Nova Publisher, 2012, 197–199.

CHAPTER 13

RESEARCH METHODOLOGIES ON PHYSICOCHEMICAL PROPERTIES AND STRUCTURE OF GRAPHITIC CARBONS

HEINRICH BADENHORST

CONTENTS

13.1   INTRODUCTION

Graphite in its various forms is a very important industrial material, it is used in a wide variety of specialized applications. These include high temperature uses where the oxidative reactivity of graphite is very important, such as electric arc furnaces and nuclear reactors. Graphite intercalation compounds are used in lithium ion batteries or as fire retardant additives. These may also be exfoliated and pressed into foils for a variety of uses including fluid seals and heat management.

Graphite and related carbon materials has been the subject of scientific investigation for longer than a century. Despite this fact there is still a fundamental issue that remains, namely the supra-molecular constitution of the various carbon materials [1, 2]. In particular it is unclear how individual crystallites of varying sizes are arranged and interlinked to form the complex microstructures and defects found in different bulk graphite materials.

Natural graphite flakes are formed under high pressure and temperature conditions during the creation of metamorphosed siliceous or calcareous sediments [3]. Synthetic graphite on the other hand is produced via a multistep, re-impregnation process resulting in very complex microstructures and porosity [4]. For both of these highly graphitic materials the layered structure of the ideal graphite crystal is well established [5]. However, in order to compare materials for a specific application the number of exposed, reactive edge sites are of great importance.

This active surface area (ASA) is critical for quantifying properties like oxidative reactivity and intercalation capacity. The ASA is directly linked to the manner in which crystalline regions within the material are arranged and interconnected. The concept of ASA has been around for a long time [6–11], however due to the nature of these sites and the very low values of the ASA for macrocrystalline graphite, it is difficult to directly measure this parameter accurately and easily. Hence it is been difficult to implement in practice and an alternative method must be employed to assess the microstructures found in graphite materials.

New developments in the field of scanning electron microscopy (SEM) allow very high resolution imaging with excellent surface definition [12, 13]. The use of high-brightness field-emission guns and in-lens detectors allow the use of very low (~ 1kV) acceleration voltages. This limits electron penetration into the sample and significantly enhancing the surface detail, which can be resolved. Due to the beneficiation and processing of the material, graphite exhibits regions of structural imperfection, which conceal the underlying microstructure. Since oxygen only gasifies graphite at exposed edges or defects, these regions may be largely removed by oxidation, leaving behind only the underlying core flake structure. This oxidative treatment will also reveal crystalline defects such as screw dislocations.

Furthermore, the oxidative reactivity of graphite is very sensitive to the presence of very low levels of impurities that are catalytically active. The catalytic activity is directly dependent on the composition of the impurity not just the individual components. Hence a pure metal will behave differently from a metal oxide, carbide or carbonate [14–16], a distinction, which is impossible to ascertain from elemental impurity analysis. The very low levels required to significantly affect oxidative properties are also close to the detection limits of most techniques, as such the only way to irrefutably verify the absence of catalytic activity is through visual inspection of the oxidized microstructure.

Thus, in combination these two techniques are an ideal tool for examining the morphology of graphite materials. Only once a comprehensive study of the microstructure of different materials has been conducted can their ASA related properties be sensibly compared.

13.2 MATERIALS AND METHODS

Four powdered graphite samples will be compared. The first two are proprietary nuclear grade graphite samples, one from a natural source (NNG) and one synthetically produced material (NSG). Both samples were intended for use in the nuclear industry and were subjected to high levels of purification including halogen treatment. The ash contents of these samples were very low, with the carbon content being >99.9 mass%. The exact histories of both materials are not known. The third graphite (RFL) was obtained from a commercial source (Graphit Kropfmühl AG Germany). This is a large flake, natural graphite powder and was purified by the supplier with an acid treatment and a high temperature soda ash burn up to a purity of 99.91 mass%. A fourth sample was produced for comparative purposes by heating the RFL sample to 2700°C for 6 hours in a TTI furnace (Model: 1000–2560–FP20). This sample was designated PRFL since the treatment was expected to further purify the material. All thermal oxidation was conducted in a TA Instruments SDT Q600 thermogravimetric analyzer (TGA) in pure oxygen. The samples were all oxidized to a burn-off of around 30%, at which point the oxidizing atmosphere was rapidly changed to inert. SEM images were obtained using an ultra-high resolution field-emission microscope (Zeiss Ultra Plus 55 FEGSEM) equipped with an in-lens detection system.

13.3 MICROSTRUCTURE OF GRAPHITE

13.3.1. FEGSEM RESOLUTION

Initially the RFL sample was only purified up to a temperature of 2400°C. When this sample was subsequently oxidized, the purification was found to have been only partially effective. It was possible to detect the effects of trace levels of catalytic impurities, as can be seen from Fig. 13.1.

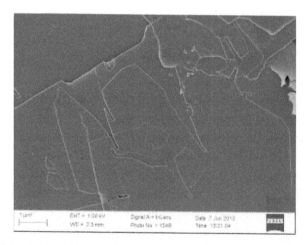

FIGURE 13.1   FEGSEM image of partially purified RFL (30k × magnification).

Since the catalyst particles tend to trace channels into the graphite, as seen in Fig. 13.2, the consequences of their presence can be easily detected. As a result it is possible to detect a single, minute catalyst particle, which is active on a large graphite flake. This effectively results in the ability to detect impurities that are present at extremely low levels.

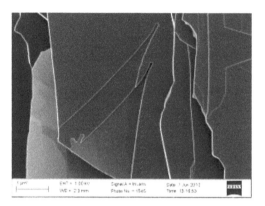

FIGURE 13.2   Channeling catalyst particles (40k × magnification).

When the tips of these channels are examined, the ability of the high resolution FEGSEM, operating at low voltages, to resolve surface detail and the presence of catalytic particles are further substantiated. As can be seen from Fig. 13.3, the microscope is capable of resolving the catalyst particle responsible for the channeling.

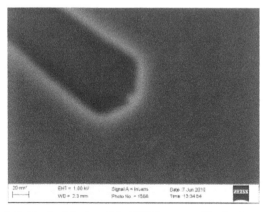

FIGURE 13.3   Individual catalyst particle (1000k × magnification).

In this case the particle in the image has a diameter of around ten nanometers. This demonstrates the powerful capability of the instrument and demonstrates its ability to detect the presence of trace impurities.

13.3.2 AS-RECEIVED MATERIAL

When the as-received natural graphite flakes are examined in Fig. 13.4, their high aspect ratio and flat basal surfaces are immediately evident.

FIGURE 13.4   Natural graphite flakes (175 × magnification).

When some particles are examined more closely, they were found to be highly agglomerated, as shown in Fig. 13.5.

FIGURE 13.5   Close-up agglomerated flake (3k × magnification).

All samples were subsequently wet-sieved in ethanol to break-up the agglomerates. When the sieved flakes are examined they are free of extraneous flakes but still appear to be composite in nature with uniform edges, as can be seen in Fig. 13.6.

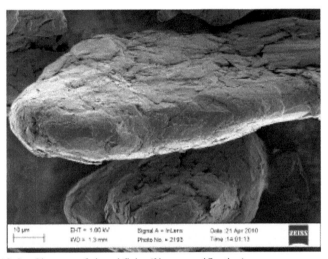

FIGURE 13.6   Close-up of sieved flake (3k × magnification).

This surface deformation is due to the beneficiation process, during which the edges tend to become smooth and rounded. As such the expected layered structure is largely obscured.

13.3.3 PURIFIED NATURAL GRAPHITE

However, when the oxidized natural graphite flakes are examined, their layered character is immediately apparent as seen in Fig. 13.7.

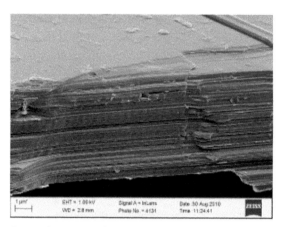

FIGURE 13.7   Layered structure of natural graphite (20k × magnification).

When the edges are examined from above, the crisp 120° angles expected for the hexagonal crystal lattice of graphite are evident as in Fig. 13.8.

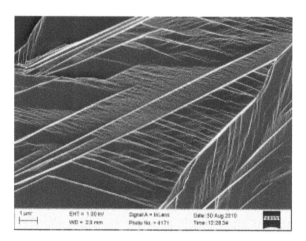

FIGURE 13.8   Hexagonal edge structures of natural graphite (20k × magnification).

The flat, linear morphology expected for a pristine graphite crystal is now more visible in Fig. 13.9.

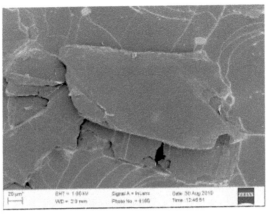

FIGURE 13.9   Oxidized natural graphite flake (800 × magnification).

When the basal surface is examined more closely as in Fig. 13.10, the surface is smooth and flat across several tens of micrometers.

FIGURE 13.10   Basal surface of natural graphite flake (3k × magnification)

Since the graphite atoms are bound in-plane by strong covalent bonding, the basal surface is expected to be comparatively inert. This surface shows no signs of direct oxidative attack, with only some minor surface steps visible. Thus the highly crystalline nature of the material is readily evident but further investigation does reveal some defects are present. Four possible defect structures are generally found in graphite [46, 47], namely:

i)   Basal dislocations
ii) Non-basal edge dislocations
iii) Prismatic screw dislocations
iv) Prismatic edge dislocations

Due to the fact that the breaking of carbon-carbon bonds is required for nonbasal dislocations, the existence of type (ii) defects is highly unlikely [17, 18]. Given the very weak van der Waals bonding between adjacent layers, however, basal dislocations of type (i) are very likely and multitudes have been documented [5, 18]. However, such defects will not be visible in the oxidized microstructure.

The next possible defect is type (iii), prismatic screw dislocations. These dislocations are easily distinguishable by the large pits that form during oxidation with a characteristic corkscrew shape [19], as visible in Fig. 13.11.

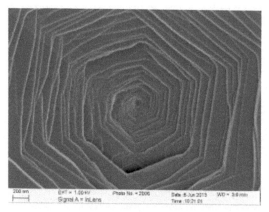

FIGURE 13.11   Prismatic screw dislocation (90k × magnification).

In general, no more than a few discrete occurrences of these defects were found in any given flake. A more prevalent defect is twinning, which is derived by a rotation of the basal plane along the armchair direction of the graphite crystal. These defects usually occur in pairs, forming the characteristic twinning band visible in Fig. 13.12.

FIGURE 13.12   Twinning band (60k × magnification).

The angled nature of these defects is more evident when they are examined edge-on as in Fig. 13.13.

FIGURE 13.13   Twinning band edge (26k × magnification).

These folds are usually caused by deformation but may also be the result of the formation process whereby impurities became trapped within the macro flake structure and were subsequently removed by purification. As is visible in the lower left hand corner of Fig. 13.13, a single sheet can undergo multiple, successive rotations and as can be seen in Fig. 13.14, the rotation angle is variable.

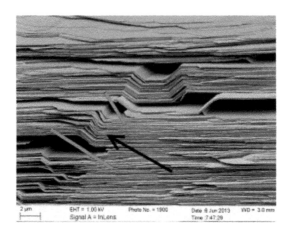

FIGURE 13.14   Rotation of twinning angle (11k × magnification).

The final defect to be considered is prismatic edge dislocations. These involve the presence of an exposed edge within the flake body, an example is shown in Fig. 13.15.

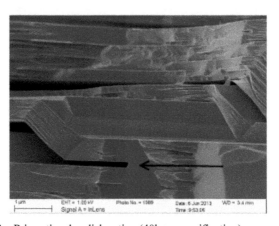

FIGURE 13.15   Prismatic edge dislocation (40k × magnification).

If the edge is small enough the structural order may be progressively restored, leading to the creation of a slit shaped pored that gradually tapers away until it disappears. An example of this behavior is demonstrated in Fig. 13.16.

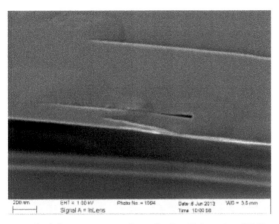

FIGURE 13.16   Gradual disappearance of small edge dislocation (125k × magnification).

If the edge dislocation is larger, when the stack collapses it will lead to a surface step, analogous to a twinning band. This can result in the formation of some very complex structures, such as the one shown in Fig. 13.17.

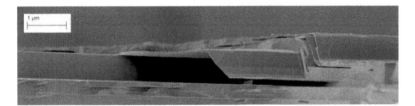

FIGURE 13.17   Complex surface structures (50k × magnification).

Thus despite being highly crystalline with an apparently straightforward geometry, complex microstructures can still be found in these natural graphite flakes.

13.3.4   CONTAMINATED NATURAL GRAPHITE

A very different microstructure is evident when the same natural graphite flakes are examined which have not been purified. Since the flakes are formed under geological processes involving high temperatures and pressures, the heat treatment step is not expected to have modified the flake microstructure. As expected the high aspect ratio and general flat shape of the flakes are still visible in Fig. 13.18.

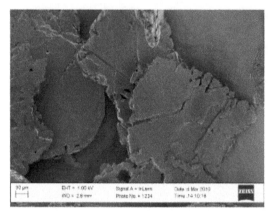

FIGURE 13.18 Flake structure of contaminated natural graphite flakes (500 × magnification).

However, when the edges of these particles are examined more closely as in Fig. 13.19, highly erratic, irregular edge features are observed.

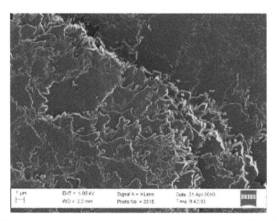

FIGURE 13.19 Erratic edge of contaminated natural graphite flakes (10k × magnification).

When the edges are scrutinized more closely, as in Fig. 13.20, the reason for these edge formations becomes clear. They are caused by minute impurities, which randomly trace channels into the graphite.

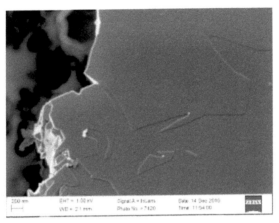

FIGURE 13.20　Catalyst activity (65k × magnification).

In certain cases, the activity is very difficult to detect, requiring the use of excessive contrast before they become noticeable as shown in Fig. 13.21.

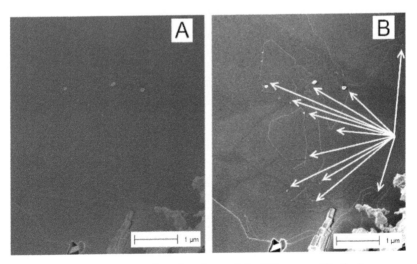

FIGURE 13.21　Contrast detection of catalyst activity (38k × magnification).

A very wide variety of catalytic behaviors were found. Broadly, these could be arranged into three categories. The first, show in Fig. 13.22, are small, roughly spherical catalyst particles. Channels resulting from these particles are in most cases triangular in nature. In general it was found that these particles tend to follow preferred channeling directions, frequently executing turns at precise, repeatable angles, as demonstrated in Fig. 13.22B. However, exceptions to these observed behaviors were also found, as illustrated in Fig. 13.22C.

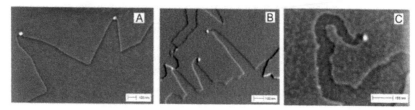

FIGURE 13.22   Small, spherical catalyst particles.

The second group contained larger, erratically shaped particles, some examples of which are shown in Fig. 13.23.

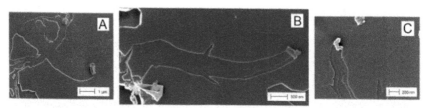

FIGURE 13.23   Small, spherical catalyst particles.

These particles exhibited random, erratic channeling. Where it is likely that the previous group may have been in the liquid phase during oxidation, this is not true for this group, since the particles are clearly capable of catalyzing channels on two distinct levels simultaneously, as can be seen in Fig. 13.23 B and C. The final group contains behaviors, which could not be easily placed into the previous two categories, of which examples are shown in Fig. 13.24.

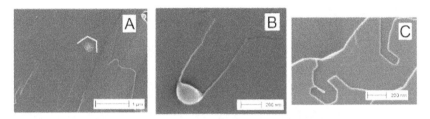

FIGURE 13.24   Small, spherical catalyst particles

The fairly large particle in Fig. 13.24A cannot be clearly distinguished as having been in the liquid phase during oxidation, yet the tip of the channel is clearly faceted with 120° angles. The particle in Fig. 13.24B was clearly molten during oxidation as it has deposited material on the channel walls. It is interesting to note that since the channel walls have expanded a negligible amount compared to the channel depth, the activity of the catalyst deposited on the wall is significantly less than that of the original particle. Finally a peculiar behavior was found in the partially purified material, where a small catalyst particle is found at the tip of a straight channel, ending in a 120° tip (clearly noticeable in Fig. 13.3). The width of the channel is roughly an order of magnitude larger than the particle itself, as seen in Fig. 13.24C. In this case channeling was always found to proceed along preferred crystallographic directions.

Such a wide variety of catalytic behaviors are not unexpected for the natural graphite samples under consideration. Despite being purified, the purification treatments are unlikely to penetrate the graphite particles completely. As such inclusions, which may have been trapped within the structure during formation, will not be removed and will be subsequently exposed by the oxidation. These impurities can have virtually any composition and hence lead to the diversity of observed behaviors.

In addition to the irregular channels, erratically shaped pits are also found in the natural graphite sample, as shown in Fig. 13.25.

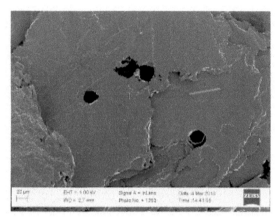

FIGURE 13.25   Pitting in natural graphite (650 × magnification).

Underdeveloped pits are often associated with erratically shaped impurities, as shown in Fig. 13.26A and 13.26B.

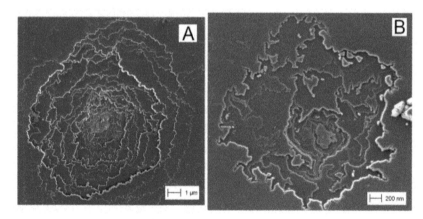

FIGURE 13.26   Impurity particles associated with pitting (35k × magnification).

The myriad of different catalytic behaviors found in this high purity natural graphite sample coupled with the enormous impact catalyst activity has on reactivity, demonstrates the danger of simply checking the impurity levels or ash content as a basis for reactivity comparison. A final morphological characteristic of this material is the presence of spiked or saw-tooth like edge formations, as can be seen in Fig. 13.27.

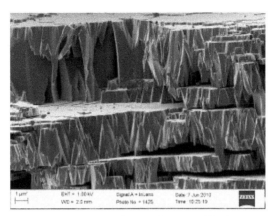

FIGURE 13.27   Saw-tooth edge formations (15k × magnification).

Closer inspection reveals that invariably the pinnacle of these struc-tures is capped by a particle, as seen in Fig. 13.28.

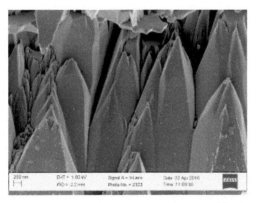

FIGURE 13.28   Close-up of saw-tooth structures (50k × magnification).

Thus these formations are caused by inactive particles, which shield the underlying graphite from attack. These layers protect subsequent lay-ers leading to the formation of pyramid like structures crowned with a single particle. In some cases as on the left hand side of Fig. 13.29, these start off as individual structures, but then as oxidation proceeds around them, the particles are progressively forced closer together to form inhibi-tion ridges, as can be observed on the right hand side of Fig. 13.29.

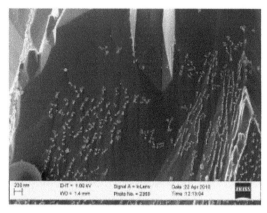

FIGURE 13.29   Inhibiting particles stacked along ridges (50k × magnification).

In extreme cases these particles may remain atop a structure until it is virtually completely reacted away, for example resulting in the nano-pyramid shown in Fig. 13.30.

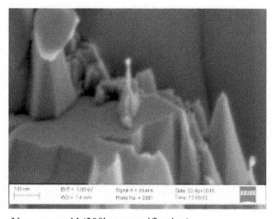

FIGURE 13.30   Nano-pyramid (300k × magnification).

In some cases particles are found which appear to neither catalyze nor inhibit the reaction, such as the spherical particles seen in Fig. 13.31.

FIGURE 13.31 Spherical edge particles (25k × magnification).

These may be catalyst particles, which agglomerate and deactivate due to their size. The graphite is oxidized away around them, until they are left at an edge, as seen in Fig. 13.32.

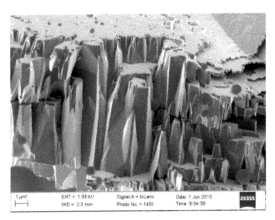

FIGURE 13.32 Spherical particles accumulating at edge (15k × magnification).

The accumulation of inhibiting particles at the graphite edge will inevitably lead to a reduction in oxidation rate as the area covered by these particles begins to constitutes a significant proportion of the total surface area. This may appreciably affect the shape of the observed conversion function, especially at high conversions.

13.3.5   NUCLEAR GRADE NATURAL GRAPHITE

The as-received nuclear grade natural graphite (NNG) exhibits a different morphology from that found in the commercial flake natural graphite. In this case the particles appear rounded and almost spherical, as shown in Fig. 13.33.

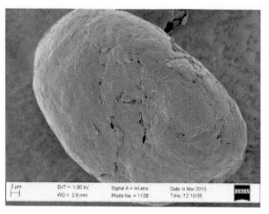

FIGURE 13.33   Rounded nuclear graphite particle (5k × magnification).

When the oxidized NNG microstructures are examined in Fig. 34, fairly complex and irregular structures are found.

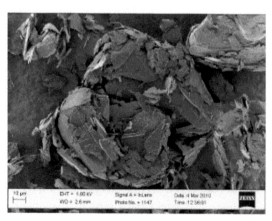

FIGURE 13.34   Oxidized NNG (1k × magnification).

The particles are extensively damaged and crumpled, however, the fact that they remain in-tact indicates that this is one continuous fragment. As the outer roughness is removed by oxidation, the multifaceted features of the particle interior are revealed. It may be concluded that these particles are in fact an extreme case of the damaged structure shown in Fig. 13.6. This material has been extensively jet-milled to create so-called "potato-shaped" graphite. Initially the particles may have resembled the commercial natural graphite flakes, however the malleability of graphite coupled with the impact deformation of jet-milling has caused them to buckle and collapse into a structure similar to a sheet of paper crumpled into a ball. Despite the high levels of purification, this material still exhibits extensive catalytic activity, similar to the flake natural graphite, as shown in Fig. 13.35.

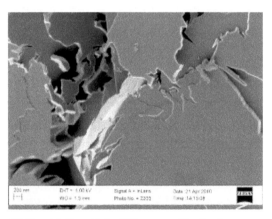

FIGURE 13.35   NNG catalytic activity (50k × magnification).

In spite of the catalytic activity and structural damage, in some regions the basal surface is still fairly smooth and flat across several micrometer, as can be seen in Fig. 13.36, indicating that the material still has good underlying crystallinity.

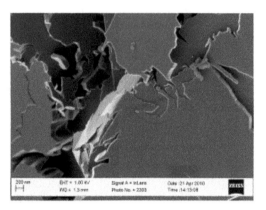

FIGURE 13.36   NNG basal plane (9k × magnification).

Thus this despite being naturally derived and evidently highly crystalline, the microstructure of the NNG material is very complex due to the extensive particle deformation during processing.

13.3.6   NUCLEAR GRADE SYNTHETIC GRAPHITE

The as-received nuclear grade synthetic graphite (NSG) exhibits a remarkably different behavior from the natural graphite samples. At first glance it is possible to distinguish between two distinct particle morphologies in Fig. 13.37.

FIGURE 13.37   Oxidized NSG (700 × magnification).

Firstly, long, thin particles are noticeable with a high aspect ratio. During the fabrication of synthetic graphite a filler material known as needle coke is used. These particles are most likely derived from the needle coke with its characteristic elongated, needle-like shape. This filler is mixed with a binder, which can be either coal tar, or petroleum derived pitch. The pitch is in a molten state when added and the mixture is then either extruded or molded. The resulting artifact can then be re-impregnated with pitch if a high density product is required. The second group of particles have a complex, very intricate microstructure and are most likely derived from this molten pitch. They are highly disordered with a characteristic mosaic texture probably derived from the flow phenomena during impregnation. When examined edge-on, the layered structure of the needle coke derived particles is still readily evident, as seen in Fig. 13.38.

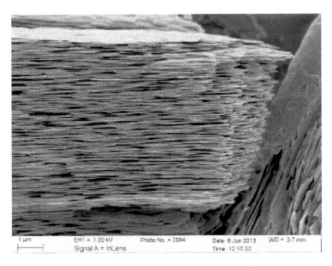

FIGURE 13.38   Oxidized NSG needle particle (20k × magnification).

The needle particles bear some resemblance to the natural graphite flakes, with the basal plane still readily identifiable in Fig. 13.39.

FIGURE 13.39   Oxidized NSG needle particle (4k × magnification).

However, when the basal plane is examined more closely in Fig. 13.40 there is a stark contrast with the natural graphite basal plane. The basal surface is severely degraded, with attack possible virtually anywhere.

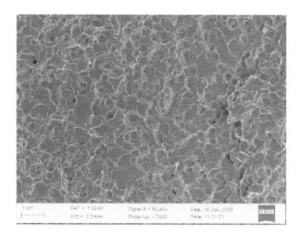

FIGURE 13.40   Oxidized NSG needle particle basal plane (25k × magnification).

The cavities were extensively investigated and no traces of impurities were found to be present. Instead the oxidation hollow has the characteristic corkscrew like shape of a screw dislocation as can be seen from Fig. 13.41. In addition, the pits tend to have a vaguely hexagonal shape.

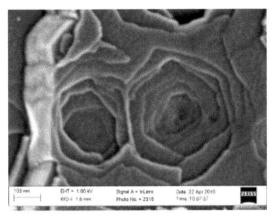

FIGURE 13.41   NSG screw dislocation (320k × magnification).

It is also important to notice that in some regions the defect density is not as high as in others, as can be seen for the different horizontal bands in Fig. 13.42A and also the different regions visible in Fig. 13.42B. This may imply different levels of crystalline perfection in these regions.

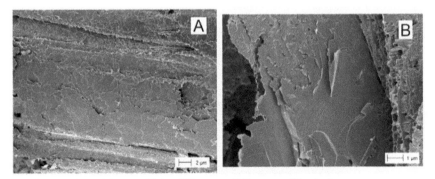

FIGURE 13.42   NSG crystallinity differences (16k × magnification).

When examined edge on as in Fig. 13.43, it can be seen that the needle particles retained their original structure, however any gaps or fissures in the folds have grown in size. This implies the development of complex slit-like porosity, probably initiated by "Mrozowski" cracks, which would not have occurred to the same extent if direct basal attack was not possible to a large degree.

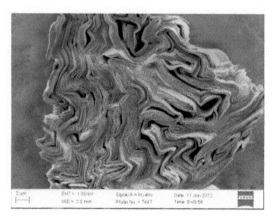

FIGURE 13.43   Slit-like pore development in NSG (8k × magnification).

When the particles edges are examined more closely in Fig. 13.44, the low level of crystalline perfection is further evident. The maximum, continuous edge widths are no more than a few hundred nanometers, far less than the several micron observable in the natural samples, such as Fig. 13.7.

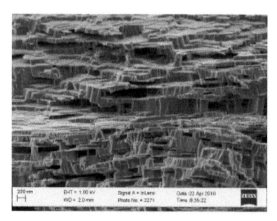

FIGURE 13.44   Degraded edge structure of NSG (50k × magnification).

The complex microstructural development characteristic of this sample is even more pronounced in the pitch particles, as can be seen from Fig. 13.45.

FIGURE 13.45   Oxidized pitch particle (4k × magnification).

These particles lack any long range order, however when their limited basal-like surfaces are examined more closely as in Fig. 13.46, a texture very similar to the basal plane of needle particles is found, indicating possibly similar levels of crystalline perfection.

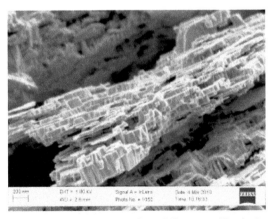

FIGURE 13.46   Oxidized pitch particle surface (88k × magnification).

On the whole, the synthetic material has the most intricate microstructural arrangement and despite the layered nature of the needle coke derived particles being readily evident, the basal surface is severely degraded indicating a high defect density. Thus this graphite can be expected to have the highest inherent ASA of all the samples considered.

13.3.7 REACTIVITY

As a comparative indication of reactivity the samples were subjected to oxidation in pure oxygen under a temperature program of 4°C/min in the TGA. The measured reaction rate as a function temperature is shown in Fig. 13.47.

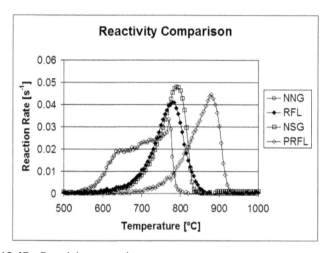

FIGURE 13.47   Reactivity comparison.

As a semiquantitative indication of relative reactivity the onset temperatures were calculated and are shown in Table 13.1.

TABLE 13.1   Onset temperatures.

	Temp (°C)
NNG	572
RFL	696
NSG	704
PRFL	760

It is clear from Table 13.1 and Fig. 13.46 that the NNG sample has the highest reactivity and PRFL the lowest. The NSG and RFL samples have similar intermediate reactivity, although the NSG sample does exhibit a higher peak reactivity. Given the microstructure and impurities found in the respective samples this result is not unexpected. The NNG and NSG samples have comparably complex microstructures, which would both have relatively high surface areas. Despite the higher crystalline perfection of the NNG sample the presence of impurities increases its reactivity significantly above that of the NSG. The sample with the lowest reactivity is the purified PRFL sample, which is not surprising since it exhibited no catalytic activity coupled with a highly crystalline structure and a flake geometry with a large aspect ratio.

The RFL material also has excellent crystallinity and a disc structure with low edge surface area. However, despite its high purity (>99.9%) the RFL sample still contains considerable amounts of catalytically active impurities, thus increasing its reactivity. It is remarkable that despite the comparatively high amount of defects and consequently high ASA, the NSG sample achieves a reactivity comparable to the RFL materials. This indicates the dramatic effect even very low concentrations of catalytically active impurities can have on the oxidation rate of graphite. As can be seen from Fig. 13.2, these minute impurities rapidly create vast amounts of additional surface area through their channeling action. This raises the ASA of the idealized, flat natural flakes to a level comparable to the synthetic material.

13.4 CONCLUSIONS

The active surface area of graphite is important for a wide variety of applications. Through the use of oxidation to expose the underlying microstructure and high-resolution surface imaging it is possible to discern between graphite materials from different origins, irrespective of their treatment histories. This establishes a direct link between the ASA based characteristics, like the oxidative reactivity, of disparate samples and their observed microstructures. This enables like-for-like comparison of materials for selection based on the specific application.

Despite a highly crystalline structure, the oxidative behavior of natural graphite can be dramatically altered through the presence of trace catalytically active impurities and structural damage induced by processing. These differences would be difficult to detect using analytical techniques such as X-ray diffraction, Raman spectroscopy or X-ray fluorescence, due to the similarity of the materials. Synthetic graphite has a much higher defect density than the natural graphite but a similar reactivity to these materials can be achieved if the material is free of catalytic impurities.

In addition, this technique enables insights regarding the extent to which the properties of different materials can be enhanced by further treatments. For example, on the basis of this investigation, it is clear that the oxidative reactivity of the NNG sample may be improved by purification, but due to the damaged structure it cannot achieve the stability observed for the PRFL material, despite both being natural graphite samples. Furthermore, despite having similar reactivities, the NSG and RFL materials have vastly different microstructures and therefore would not be equally suitable for applications where, for example, inherent surface area is very important. In conclusion, given the complexity found in different graphite materials, it is critical that the microstructure should be considered in conjunction with kinetic and other ASA related parameters to afford a comprehensive understanding of the material properties.

This is by no means an exhaustive study of all possible morphologies found in natural and synthetic graphite materials but it does demonstrate some of the intricate structures that are possible.

KEYWORDS

- **graphitic carbons**
- **micro structures**
- **physicochemical properties**
- **Scanning electron microscopy (SEM)**
- **surface area**
- **synthetic material**

REFERENCES

1. Radovich, L. R., Physicochemical properties of carbon materials: a brief overview. In: Serp P, Figueiredo, J. L., editors. Carbon materials for catalysis, Hoboken, NJ: Wiley; 2009, p. 1–34.
2. Harris, P. J. F., New perspectives on the structure of graphitic carbons. Crit Rev Solid State Mater Sci 2005; 30: 235–53.
3. Luque, F. J., Pasteris, J. D., Wopenka B, Rodas M, Barranechea, J. F., Natural fluid-deposited graphite: mineralogical characteristics and mechanisms of formation. American Journal of Science. 1998; 298: 471–98.
4. Pierson, H. O., Handbook of carbon, graphite, diamond and fullerenes. Properties, processing and applications. New Jersey, USA: Noyes Publications; 1993.
5. Reynolds, W. N., Physical Properties of Graphite. Amsterdam: Elsevier; 1968.
6. Laine, N. R., Vastola, F. J., Walker, P. L., Importance of active surface area in the carbon-oxygen reaction. Journal of Physical Chemistry. 1963; 67: 2030–4.
7. Thomas, J. M., Topographical studies of oxidized graphite surfaces: a summary of the present position. Carbon. 1969; 7: 350–64.
8. Bansal, R. C., Vastola, F. J., Walker, P. L., Studies on ultra-clean carbon surfaces, I. I. I., Kinetics od chemisorption of hydrogen on graphon. Carbon. 1971; 9: 185–92.
9. Radovic, L. R., Walker, P. L., R. G. J. Importance of carbon active sites in the gasification of coal chars. Fuel. 1983; 62: 849–56.
10. Walker, P. L., R. L. J, J. M. T. An update on the carbon-oxygen reaction. Carbon. 1991; 29: 411–21.
11. Arenillas A, Rubiera F, Pevida C, Ania, C. O., Pis, J. J., Relationship between structure and reactivity of carbonaceous materials. Journal of Thermal Analysis and Calorimetry. 2004; 76: 593–602.
12. Cazaux, J. From the physics of secondary electron emission to image contrasts in scanning electron microscopy. J Electron Microsc (Tokyo). 2012; 61(5): 261–84.
13. Lui, J. The Versatile, F. E. G.,-SEM: From Ultra-High Resolution To Ultra-High Surface Sensitivity. Microscopy and Microanalysis. 2008; 9: 144–5.
14. Baker, R. T. K., Factors controlling the mode by which a catalyst operates in the graphite-oxygen reaction. Carbon. 1986; 24: 715–7.
15. Yang, R. T., Wong, C. Catalysis of carbon oxidation by transition metal carbides and oxides. Journal of Catalysis. 1984; 85: 154–68.
16. McKee, D. W., Chatterji, D. The catalytic behavior of alkali metal carbonates and oxides in graphite oxidation reactions. Carbon. 1975; 13: 381–90.
17. Fujita, F. E., Izui K. Observation of lattice defects in graphite by electron microscopy, Part 1. J Phys Soc Japan 1961; 16(2): 214–7.
18. Suarez-Martinez, I, Savini, G, Haffenden, G, Campanera, J. M., Heggie, M. I., Dislocations of Burger's Vector $c/2$ in graphite. Phys Status Solidi C 2007; 4(8): 2958–62.
19. Rakovan, J, Jaszczak, J. A., Multiple length scale growth spirals on metamorphic graphite {001} surfaces studied by atomic force microscopy. American Mineralogist. 2002; 87: 17–24.

CHAPTER 14

THE EFFECT OF ANTIOXIDANT COMPOUNDS ON OXIDATIVE STRESS IN UNICELLULAR AQUATIC ORGANISMS

O. V. KARPUKHINA, K. Z. GUMARGALIEVA,
and A. N. INOZEMTSEV

CONTENTS

14.1   INTRODUCTION

Toxic effect of heavy metal salts and hydrogen peroxide—lipid peroxidation inducers—was studied on the culture of *Paramecium caudatum* cells. A significant protective effect of substances with antioxidant effect (ascorbic acid, piracetam) in the experimental model of induced oxidative stress in unicellular infusoria is established.

Heavy metals are potentially dangerous toxicants, which are the destabilizing factor in the ecological system of established biocoenosis. lead, cadmium, mercury, cobalt, and other heavy metals may be inducers of the oxidative stress, which is based on formation of excessive quantity of free radicals [1, 2]. Their reactivity is extremely high and initiates chain oxidation reaction. Free radicals become the reason for serious functional disorders, because various cell components are damaged [3]. Initiation of lipid peroxidation in biological membranes is an example and promotes disturbance of their structure and penetrability increase. Specialized enzymatic and nonenzymatic antioxidant systems are protection against free radicals [4].

Determination of substances which action is aimed at normalizing metabolic processes, blockage of pathological free-radical processes is the important branch of investigations of organism adaptation to the impact of toxic agents. It is common knowledge that such compound is ascorbic acid able to react with superoxide and hydroxyl radicals and thus decreases their concentration in the cell, preventing development of the oxidant stress[5]. In tests on rats, it was found that ascorbic acid caused protective antioxidant effect on formation of conditioned responses under toxic effect of joint action of heavy metal salts (Co, Pb, and Cd) and neurotropic preparation piracetam [6, 8]. It has been found that catalytic action of heavy metal salts affects piracetam structural characteristics and, consequently, distorts its curative effect [7]. It is shown in vitro that ascorbic acid prevents heavy metal salts interaction with the drug preparation structure, and injection of the antioxidant into animals has eliminated the negative effect of heavy metals on formation of adaptive reactions[8].

The complex organism responses which *is protective and* provides adaptation to varying conditions in response to the impact of various adverse factors, including toxic agents, is based on global fundamental mechanisms. Unicellular infusoria may be considered as simple receptor-effector systems capable of rapidly respond to a chemical action by entire set of biological, physiological and biochemical changes: chemotaxis, backward mutation of the ciliate activity, the rate of reproduction, and phagocytic activity. The aim of this work was to study the effect of toxic agents, the inductors of oxidative stress and protective action of antioxidant agents, on the vital activity of unicellular aquatic organisms *Paramecium caudatum.*

14.2 EXPERIMENTAL

Paramecium caudatum (the wild strain) cell culture was cultivated in the Lozina-Lozinsky medium with addition of nutritive medium yeasts containing *Saccharomyces cerevisiae* yeasts. The oxidative stress was induced by: cadmium chloride, lead acetate, and hydrogen peroxide (H_2O_2). Infusorian sensitivity to toxic agents was determined by time of their death established by protozoa motion cessation, often accompanied by cell deformation and cytolysis. The exposure time was 2 hours. The control in all tests was the number of cells in 10 mL of the medium containing the intact culture of infusorian (without oxidative stress induction).

Total number of cells in 10 mL of the medium containing infusorian was determined in the Goryaev counting chamber. Cells sampled at the stationary phase of growth were incubated in the medium with the chemical substance added at $(20 \pm 2)°C$ temperature during 15 days (chronic effect).

The effect of toxic agents was studied in several concentrations (0.05, 0.025, 0.005, 0.0025, and 0.00125 mg/L); the effect of substances with antioxidant properties (ascorbic acid, piracetam) was studied at concentrations of 1, 10, 50 µM, and 1 mM.

14.3 RESULTS AND DISCUSSION

As established in the experiments performed, in the initial 30 min of incubation, cadmium chloride and lead acetate (0.05 mg/L) caused immediate motion cessation of infusorian with the shape change to rod that showed structural changes in the cell membranes. Hydrogen peroxide (0.05 mg/L) initiated degradation of the lipid part of the unicellular organism membrane: gross morphological changes in the cellular membranes of infusorian as multiple subglobose projections (the cell membrane "vacuolization") occurred, which then broke and caused death of the cell-organism. The number of species in control does not change during the entire test (Fig. 14.1). Hydrogen peroxide injection to the medium with infusorian caused 3-fold decrease of the number of species in the first minute of the test. Decrease in population of test cells indicates the destructive membrane pathology in which occurrence free-radical oxidation processes are of importance. By 60th minute of the test, 100% infusorian death was observed. The toxic effect of heavy metal salts was less expressed.

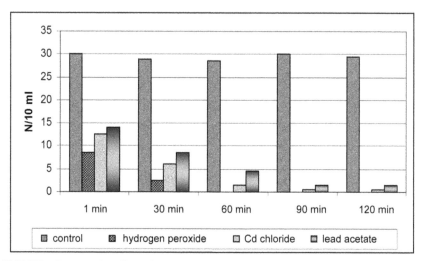

FIGURE 14.1 Toxic effect of chemical agents on *Paramecium caudatum* cell culture (abscissa axis—test duration; axis of ordinates—the number of species in 10 mL of the test medium).

The goal of the following test was the study of hydrogen peroxide and heavy metal salts effect on reproduction of Paramecium cells. Since the cell culture death at 0.05 and 0.025 mg/L concentration was observed (Fig. 14.1), concentrations of toxic agents in tests on the cell reproduction were 20-fold decreased. At this concentration, a considerable part of cells survived during the first 24 hours of the test (Fig. 14.2). For the cell incubation during 15 days, media with hydrogen peroxide and cadmium chloride, in which infusorian have not reproduced even the first generation, were found the most toxic ones (Fig. 14.2).

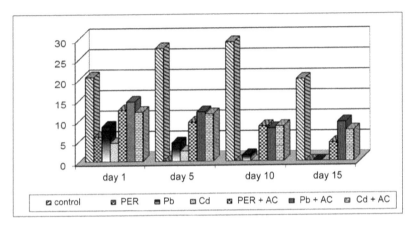

FIGURE 14.2   Dynamics of *Paramecium caudatum* population changes in the medium containing various chemical agents: control—normal saline; PER—hydrogen peroxide; Pb—lead diacetate; AC—ascorbic acid.

By the 5th day, infusorian incubated in hydrogen peroxide died. Adding of ascorbic acid (1 µM) to the cell medium subjected to hydrogen peroxide action almost two-fold increased the number of species in the first 24 hours. At the same time, ascorbic acid prevented death of culture cells that was noted in all test days (Figs. 14.2, and 14.3).

In the medium with a heavy metal salt added, 100% death rate of the culture was observed on the 15th day.

By the 15th day, only an insignificant decrease in the species population was observed, when ascorbic acid or lead acetate was added to the cell culture. Thus, here ascorbic acid manifested similar antioxidant activity as against hydrogen peroxide.

To prove efficiency of an antioxidant agent with the action mechanism different from ascorbic acid, we have studied Paramecium caudatum in the medium with added piracetam—the compound with nonspecific antioxidant activity [9] (Fig. 14.3).

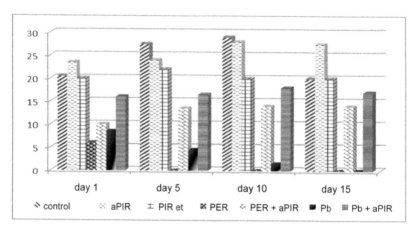

FIGURE 14.3 Dynamics of *Paramecium caudatum* population changes in the medium containing various chemical agents: control—normal saline; PER—hydrogen peroxide; Pb—lead diacetate; AC—ascorbic acid; PIR et—piracetam solution (drag form); aPIR—structurally modified piracetam solution.

In these tests, the efficiency of the piracetam drug form (PIR et) and its structurally modified form (aPIR). Both substances prevented death of infusorian cells due to hydrogen peroxide and heavy metal salt action. The antioxidant effect of piracetam was less expressed as compared with protective action of ascorbic acid. Nevertheless, by the 15th day survivability of species with injection of aPIR into the medium with infusorian reached 100%.

14.4 CONCLUSION

The experiments performed have shown that responses of unicellular aquatic organisms are effective for obtaining primary information on cytotoxicity of substances (heavy metal salts, peroxides, etc.).

Ascorbic acid and membrane-stabilizing antioxidant substance—piracetam significantly increase cell nonresponsiveness to heavy metals that proves the important role of antioxidant compounds in free-radical oxidation processes in the organism cell.

KEYWORDS

- antioxidant compounds
- heavy metals
- lead, cadmium, mercury, cobalt
- oxidative stress
- paramecium caudatum cells
- unicellular aquatic organisms

REFERENCES

1. Simmons, So, Fan, Cy. Ramabhadran, R. Cellular stress response pathway system as a sentinel ensemble in toxicological screening. Toxically Sci. 2009, 111(2): 202.
2. Leonard, S. S., Harris, G. K., Shi, X. L. Metal-induced oxidative stress and signal transduction. Free Rad Boil Med., 2004, 37(12): 1921.
3. Pryor, W. A. Oxy-Radicals and Related Species: Their Formation, Lifetimes and Reactions. Annual Review of Physiology. 1986, 48. 657.
4. Flora, S. J. Structural, chemical and biological aspects of antioxidants for strategies against metal and metalloid exposure. Oxid Med Cell. 2009, 2(4): 191.
5. Davies, M., Austin, J., Partridge, D. Vitamin C: Its Chemistry and Biochemistry, Royal Society of Chemistry. London. UK. 1991.
6. Bokieva, S. B., Karpukhina, O. V., Gumargalieva, K. Z., Inozemtsev, A. N. Combined effects of heavy metals and Piracetam destroying the adaptive behavior formation. Proceedings of Gorsky Stat Univ of Agricul. 2012, 4(4): 194.
7. Karpukhina, O. V., Gumargalieva, K. Z., Soloviev, A. G., Inozemtsev, A. N. Effects of lead diacetate on structure transformation and functional properties of piracetam. Journal of Environmental Protection and Ecology. 2004, 5 (3): 577.
8. Karpukhina, O. V., Gumargalieva, K. Z., Bokieva, S. B., Kalyuzhny, A. L., Inozemtsev, A. N. Ascorbic acid protects the body from toxic effects of cadmium. "High Technology, basic and applied researches in physiology and medicine" Saint-Petersburg, Russia. 2011, 3(4): 166.
9. Gouliaev, A. H., Senning, A. Piracetam and other structurally related no tropics. Brain Res. Rev. 1994, 19(2): 180.

CHAPTER 15

A LECTURE NOTE ON DETERMINATION OF ACID FORCE OF COMPONENTS OF SYNTHESIS OF 1-[2-(O-ACETYLMETHYL)-3-O-ACETYL-2-ETHYL]-METHYLDICHLORINEPHOSPHITE

V. A. BABKIN, V. U. DMITRIEV, G. A. SAVIN, E. S. TITOVA, and G. E. ZAIKOV

CONTENTS

15.1   INTRODUCTION

Quantum-chemical calculation of molecule of 1-[2-(o-acetylmethyl)-3-o-acetyl-2-ethyl]-methyldichlorinephosphite (I) and components of its synthesis [4] for the first time is executed by method AB INITIO in basis 6–311G**. The optimized geometrical and electronic structures of these compounds are received. Their acid force is theorized. All these compounds concern to a class of very weak C–H-acids (pKa > 14, where pKa is the universal parameter of acidity). Dependence between acid force of components of synthesis pKa (II) and pKa (III) and acid force pKa (I) a required product is positioned:

$$pKa(I) = pKa \ (II) + \sqrt{\frac{pKa(III)}{4}} \ .$$

Acetyloxymethyl-2-chlorine-5-ethyl-1,2,3-dioxaphospho-rynan, acetyl chloride, 1-[2-(o-acetylmethyl)-3-o-acetyl-2-ethyl]-methyldichlorinephosphite 1-[2-(o-Acetylmethyl)-3-o-acetyl-2-ethyl]-methyldichlorinephosphite (I) is intermediate substance for reception of a medical product from a hepatitis. It possesses interesting and probably unique properties.

This compound is received at interaction of acetyl chloride (III) and 5-acetyloxymethyl-2-chlorine-5-ethyl-1,2,3-dioxaphosphorynan (II) in a gas phase:

The Mechanism of reaction of acylation of bicyclophosphites by chlorine anhydride of carboxylic acids consists of three stages. The first stage of reaction is investigated in Ref. [4]. The mechanism of the second stage of reaction now is not studied. One of the first investigation phases of the mechanism of synthesis of studied compound (I) can consider an estimation of acid force of components of synthesis pKa (II) and pKa (III) and a required product pKa (I) and an establishment of dependence between them.

The purpose of the present work was quantum-chemical calculation of components of synthesis of acetyl chloride (III) and 5-acetyloxy-methyl-2-chlorine-5-ethyl-1,2,3-dioxaphosphorynan (II) and a product 1-[2-(o-acetylmethyl)-3-o-acetyl-2-ethyl]-methyldichlorinephosphite (I) by method AB INITIO in basis 6–311G**, a theoretical estimation of their acid force and an establishment of dependence between acid force of components of synthesis pKa (III) and pKa (II) and pKa (I).

15.2 EXPERIMENTAL

Method AB INITIO in basis 6–311G** with optimization of geom-etry on all parameters by the standard gradient method which has been built in in PC GAMESS, used for quantum-chemical calculation of components of synthesis of 1-[2-(o-acetylmethyl)-3-o-acetyl-2-ethyl]-methyldichlorinephosphite (I) [1]. Calculation was carried out in approach of the isolated molecule a gas phase. The theoretical estimation of acid force of components of synthesis was carried out under the formula [2]:

$$pKa = 49.04 - 134.61 \times q_{MAX}^{H+},$$

where pKa is a universal parameter of acidity, and q_{MAX}^{H+} is the maximal charge on atom of hydrogen of a molecule.

For visual representation of models of components of synthesis pro-gram MasMolPlt [3] was used.

15.3 RESULTS OF CALCULATIONS

The Optimized geometrical and electronic structures, the general en-ergy, electronic energy, lengths of bonds and valent corners of product 1-[2-(o-acetylmethyl)-3-o-acetyl-2-ethyl]-methyldichlorinephosphite (I) are received by method AB INITIO in basis 6–311G** and presented on Fig. 15.1 and in Table 15.1. Values pKa of all components of synthesis are certain by means of the formula, which used with success in Refs. [5–24]:

$$pKa = 49.04 - 134.61 \times q_{MAX}^{H+}$$

$q_{MAX}^{H^+}(I) = +0.14$, $q_{MAX}^{H^+}(II) = +0.16$ and $q_{MAX}^{H^+}(III) = +0.14$. Accordingly, pKa (I) = 30.2, pKa (II) = 27.5 and pKa (III) = 30.1. By us it has been positioned, that between acid force of components of synthesis pKa (II) and pKa (III) and acid force of a received product pKa (I) there is a following dependence:

$$pKa(I) = pKa\ (II) + \sqrt{\frac{pKa(III)}{4}}$$

The General energy (E_0), the electronic energy (E_{el}), the maximal charge on atom of hydrogen—$q_{MAX}^{H^+}$ and values pKa of components of synthesis pKa (II) and pKa (III) and a product pKa (I) are presented in Table 15.2.

Quantum-chemical calculation of a molecule 1-[2-(o-acetylmethyl)-3-o-acetyl-2-ethyl]-methyldichlorinephosphite (I) for the first time is executed by method AB INITIO in basis 6–311G**. Quantum-chemical calculation of components of synthesis (II) and (III) has been executed by us [4].

The optimized geometrical and electronic structures of compound (I) is received. Acid force is theorized. By us it is shown, that all of these compounds concern to a class weak C-H-acids (pKa > 14). Dependence between acid force of components of synthesis (II) and (III) and also acid force of a product (I) is positioned.

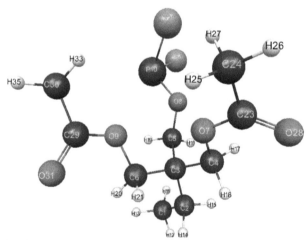

FIGURE 15.1   Geometric and electronic structure of a molecule of 1-[2-(o-acetylmethyl)-3-o-acetyl-2-ethyl]-methyldichlorinephosphite ($E_0 = -5,306,055$ kDg/mol, $E_{el} = -10,395,257$ kDg/mol).

TABLE 15.1 Optimized lengths of bonds, valency corners and charges of atoms of a molecule of 1-[2-(o-acetylmethyl)-3-o-acetyl-2-ethyl]-methyldichlorinephosphite.

Bond lengths	R, Å	Valency corners	Grad	Atom	Charge (by Milliken)
C(1)-C(2)	1.53			C1	-0.23
C(2)-C(3)	1.55	C(3)C(2)C(1)	117	C2	-0.21
C(3)-C(4)	1.53	C(4)C(3)C(2)	105	C3	-0.37
C(3)-C(5)	1.53	C(5)C(3)C(2)	108	C4	0.19
C(3)-C(6)	1.53	C(6)C(3)C(2)	107	C5	0.17
C(4)-O(7)	1.41	O(7)C(4)C(3)	111	C6	0.19
C(5)-O(8)	1.42	O(8)C(5)C(3)	111	O7	-0.44
C(6)-O(9)	1.41	O(9)C(6)C(3)	110	O8	-0.66
O(8)-P(10)	1.58	P(10)O(8)C(5)	121	O9	-0.45
C(1)-H(11)	1.08	H(11)C(1)C(2)	112	P10	0.85
C(1)-H(12)	1.08	H(12)C(1)C(2)	107	H11	0.09
C(1)-H(13)	1.08	H(13)C(1)C(2)	107	H12	0.11
C(2)-H(14)	1.08	H(14)C(2)C(3)	108	H13	0.10
C(2)-H(15)	1.08	H(15)C(2)H(14)	105	H14	0.11
C(4)-H(16)	1.08	H(16)C(4)O(7)	108	H15	0.11
C(4)-H(17)	1.08	H(17)C(4)H(16)	107	H16	0.11
C(5)-H(18)	1.08	H(18)C(5)O(8)	106	H17	0.12
C(5)-H(19)	1.08	H(19)C(5)H(18)	108	H18	0.13
C(6)-H(20)	1.08	H(20)C(6)O(9)	107	H19	0.11
C(6)-H(21)	1.08	H(21)C(6)H(20)	107	H20	0.11
P(10)-Cl(22)	2.06	Cl(22)P(10)O(8)	98	H21	0.12
O(7)-C(23)	1.32	C(23)O(7)C(4)	116	Cl22	-0.26
C(23)-C(24)	1.50	C(24)C(23)O(7)	111	C23	0.47
C(24)-H(25)	1.08	H(25)C(24)C(23)	109	C24	-0.25
C(24)-H(26)	1.08	H(26)C(24)C(25)	110	H25	0.12
C(24)-H(27)	1.08	H(27)C(24)C(25)	108	H26	0.13
C(23)-O(28)	1.18	O(28)C(23)O(7)	123	H27	0.14
O(9)-C(29)	1.32	C(29)O(9)C(6)	116	O28	-0.47
C(29)-C(30)	1.50	C(30)C(29)O(9)	111	C29	0.47
C(29)-O(31)	1.18	O(31)C(29)O(9)	123	C30	-0.26
P(10)-Cl(32)	2.08	Cl(32)P(10)O(8)	100	O31	-0.46
C(30)-H(33)	1.08	H(33)C(30)C(29)	109	Cl32	-0.29
C(30)-H(34)	1.08	H(34)C(30)H(33)	107	H33	0.13
C(30)-H(35)	1.08	H(35)C(30)H(34)	110	H34	0.13
				H35	0.13

TABLE 15.2 The general energy, energy of bonds, the maximal charge on atom of hydrogen, a universal parameter of acidity of components of synthesis of 1-[2-(o-acetylmethyl)-3-o-acetyl-2-ethyl]-methyldichlorinephosphite.

No	The Component of synthesis	E_0, kDg/mol	E_{el}, kDg/mol	q_{MAX}^{H+}	pKa
1.	5-Acetyloxymethyl-2-chlorine-5-ethyl-1,2,3-dioxaphosphorynan	-3,700,998	-7,080,777	+0.16	27.5
2.		-5,306,055	-10,395,257	+0.14	30.2
3.	1-[2-(o-Acetylmethyl)-3-o-acetyl-2-ethyl]-methyldichlorinephosphite	-1,604,983	-1,998,646	+0.14	30.1
	Acetyl chloride				

REFERENCES

1. Schmidt, M. W., Baldrosge, K. K., Elbert, J. A., Gordon, M. S., Enseh, J. H., Koseki, S., Matsvnaga, N., Nguyen, K. A., Su, S. J. et al. *J. Computer Chem.* 14, 1347–1363 (1993).
2. Babkin, V. A., Fedunov, R. G., Minsker, K. S., et al. *Oxidation communication.* 2002. № 1, 25, 21–47.
3. Bode, B. M., Gordon, M. S. *J. Mol. Graphics Mod.* 16, 1998, 133–138.
4. Babkin, V. A., Dmitriev, V. U., Savin, G. A., Zaikov, G. E., Estimation of acid force of components of synthesis of 5-acetyloxymethyl-2-chlorine-5-ethyl-1,2,3-dioxaphosphorynan. Moscow. *Encyclopedia of the engineer-chemist*, 3, 2009, 11–13.
5. Babkin, V. A., Fedunov, R. G., Ostrouhov, A. A., Kudryashov, A. V., Estimation of acid force of molecule 6- methilperhydrotetraline. In book: Quantum chemical calculation of unique molecular system. Vol. II. Publisher VolSU, c. Volgograd, 2010, 74–77.
6. Babkin, V. A., Fedunov, R. G., Ostrouhov, A. A., R. A. Reshetnikov. Estimation of acid force of molecule 7- methilperhydrotetraline. In book: Quantum chemical calculation of unique molecular system. Vol. II. Publisher VolSU, c. Volgograd, 2010, 77–80.
7. Babkin, V. A., Fedunov, R. G., A. A. Ostrouhov. About geometrical and electronic structure of molecule gopan. In book: Quantum chemical calculation of unique molecular system. Vol. II. Publisher VolSU, c. Volgograd, 2010, 80–83.
8. Babkin, V. A., Fedunov, R. G., A. A. Ostrouhov. About geometrical and electronic structure of molecule diagopan. In book: Quantum chemical calculation of unique molecular system. Vol. II. Publisher VolSU, c. Volgograd, 2010, 83–88.
9. Babkin, V. A., Dmitriev, V. U., Zaikov, G. E., Quantum chemical calculation of molecule DDT. In book: Quantum chemical calculation of unique molecular system. Vol. II. Publisher VolSU, c. Volgograd, 2010, 13–16.
10. Babkin, V. A., Andreev, D. S. Quantum chemical calculation of molecule 3-methylcyclopentene by method AB INITIO. In book: Quantum chemical calculation of unique molecular system. Vol. II. Publisher VolSU, c. Volgograd, 2010, 101–103.

11. Babkin, V. A., Andreev, D. S. Quantum chemical calculation of molecule cyclopentene by method AB INITIO. In book: Quantum chemical calculation of unique molecular system. Vol. II. Publisher VolSU, c. Volgograd, 2010, 103–105.

12. Babkin, V. A., Andreev, D. S. Quantum chemical calculation of molecule metylencyclobutane by method AB INITIO. In book: Quantum chemical calculation of unique molecular system. Vol. II. Publisher VolSU, c. Volgograd, 2010, 105–107.

13. Babkin, V. A., Andreev, D. S. Quantum chemical calculation of molecule 1,2-dicyclopropylethylene by method AB INITIO. In book: Quantum chemical calculation of unique molecular system. Vol. II. Publisher VolSU, c. Volgograd, 2010, 108–110.

14. Babkin, V. A., Andreev, D. S. Quantum chemical calculation of molecule isopropenylcyclopropane by method AB INITIO. In book: Quantum chemical calculation of unique molecular system. Vol. II. Publisher VolSU, c. Volgograd, 2010, 110–112.

15. Babkin, V. A., Dmitriev, V. Yu., Zaikov, G. E. Quantum chemical calculation of molecule heterolytic base uracyl. In book: Quantum chemical calculation of unique molecular system. Vol I. Publisher VolSU, c. Volgograd, 2010, 43–45.

16. Babkin, V. A., Dmitriev, V. Yu., Zaikov, G. E. Quantum chemical calculation of molecule heterolytic base adenin. In book: Quantum chemical calculation of unique molecular system. Vol I. Publisher VolSU, c. Volgograd, 2010, 45–47.

17. Babkin, V. A., Dmitriev, V. Yu., Zaikov, G. E. Quantum chemical calculation of molecule heterolytic base guanin. In book: Quantum chemical calculation of unique molecular system. Vol I. Publisher VolSU, c. Volgograd, 2010, 47–49.

18. Babkin, V. A., Dmitriev, V. Yu., Zaikov, G. E. Quantum chemical calculation of molecule heterolytic base timin. In book: Quantum chemical calculation of unique molecular system. Vol I. Publisher VolSU, c. Volgograd, 2010, 49–51.

19. Babkin, V. A., Dmitriev, V. Yu., Zaikov, G. E. Quantum chemical calculation of molecule heterolytic base cytozin. In book: Quantum chemical calculation of unique molecular system. Vol I. Publisher VolSU, c. Volgograd, 2010, 51–53.

20. Babkin, V. A., Andreev, D. S. Quantum chemical calculation of molecule isobutilene by method AB INITIO. In book: Quantum chemical calculation of unique molecular system. Vol I. Publisher VolSU, c. Volgograd, 2010, 157–159.

21. Babkin, V. A., Andreev, D. S. Quantum chemical calculation of molecule 2-methylenbutene-1 by method AB INITIO. In book: Quantum chemical calculation of unique molecular system. Vol I. Publisher VolSU, c. Volgograd, 2010, 159–161.

22. Babkin, V. A., Andreev, D. S. Quantum chemical calculation of molecule 2-methylbutene-2 by method AB INITIO. In book: Quantum chemical calculation of unique molecular system. Vol I. Publisher VolSU, c. Volgograd, 2010, 161–162.

23. Babkin, V. A., Andreev, D. S. Quantum chemical calculation of molecule 2-methylpentene-1 by method AB INITIO. In book: Quantum chemical calculation of unique molecular system. Vol I. Publisher VolSU, c. Volgograd, 2010, 162–164.

24. Babkin, V. A., Andreev, D. S. Quantum chemical calculation of molecule 2-ethylbutene-1 by method AB INITIO. In book: Quantum chemical calculation of unique molecular system. Vol I. Publisher VolSU, c. Volgograd, 2010, 164–166.

CHAPTER 16

RESEARCH METHODOLOGY ON DESIGN AND SYNTHESIS OF HYDROGEL-BASED SUPPORTS

D. HORÁK and H. HLHDKOVÁ

CONTENTS

16.1 INTRODUCTION

Superporous poly(2-hydroxyethyl methacrylate) (PHEMA) supports with pore size from tens to hundreds micrometers were prepared by radical polymerization of 2-hydroxyethyl methacrylate (HEMA) with 2 wt.% eth-ylene dimethacrylate (EDMA) with the aim to obtain a support for cell cultivation. Superpores were created by the salt-leaching technique using NaCl or $(NH_4)_2SO_4$ as a porogen. Addition of liquid porogen (cyclohexa-nol/dodecan-1-ol (CyOH/DOH) = 9/1 w/w) to the polymerization mixture did not considerably affect formation of meso- and macropores. The pre-pared scaffolds were characterized by several methods including water and cyclohexane regain by centrifugation, water regain by suction, scanning electron microscopy (SEM), mercury porosimetry and dynamic desorp-tion of nitrogen. High-vacuum scanning electron microscopy (HVSEM) confirmed permeability of hydrogels to 8-μm microspheres, whereas low-vacuum scanning electron microscopy (LVSEM) at cryo-conditions showed the undeformed structure of the frozen hydrogels. Interconnection of pores in the PHEMA scaffolds was proved. Water regain determined by centrifugation method did not include volume of large superpores (im-prints of porogen crystals), in contrast to water regain by suction method. The porosities of the constructs ranging from 81 to 91% were proportional to the volume of porogen in the feed.

Polymer supports have received much attention as microenvironment for cell adhesion, proliferation, migration and differentiation in tissue engi-neering and regenerative medicine. The three-dimensional scaffold struc-ture provides support for high level of tissue organization and remodeling. Regeneration of different tissues, such as bone [1], cartilage [2], skin [3], nerves [4], or blood vessels [5] is investigated using such constructs. An ideal polymer scaffold should thus mimic the living tissue, that is, possess a high water content, with possibility to incorporate bioactive molecules allowing a better control of cell differentiation. At the same time it re-quires a range of properties including biocompatibility and/or biodegrad-ability, highly porous structure with communicating pores allowing high cell adhesion and tissue in-growth. The material should be sterilizable and also possess good mechanical strength. Both natural and synthetic hydrogels are being developed. The advantage of synthetic polymer matrices consists in their easy proccessability, tunable physical and chemical properties, susceptibility to modifications and possibly controlled degradation.

Many techniques have been developed to fabricate highly porous constructs for tissue engineering. They include for instance solvent casting [6], gas foaming [7] or/and salt leaching [8], freeze-thaw procedure [9, 10], supercritical fluid technology [11] (disks exposed to CO_2 at high pressure) and electrospinning (for nanofiber matrices) [12]. A wide range of polymers was suggested for scaffolds. In addition to natural materials, such as collagen, gelatin, dextran [13], chitosan [14], phosphorylcholine [15], alginic [16] and hyaluronic acids, it includes also synthetic polymers, e.g., poly(vinyl alcohol) [17], poly(lactic acid) [1, 18], polycaprolactone [19], poly(ethylene glycol) [20], polyacrylamide [21], polyphosphazenes [22], as well as polyurethane [23].

Among various kinds of materials being used in biomedical and pharmaceutical applications, hydrogels composed of hydrophilic polymers or copolymers find a unique place. They have a highly water-swollen rubbery three-dimensional structure, which is similar to natural tissue [24, 25]. In this report, poly(2-hydroxyethyl methacrylate) (PHEMA) was selected as a suitable hydrogel intended for cell cultivation. The presence of hydroxyl and carboxyl groups makes this polymer compatible with water, whereas the hydrophobic methyl groups and backbone impart hydrolytic stability to the polymer and support the mechanical strength of the matrix [26]. PHEMA hydrogels are known for their resistance to high temperatures, acid and alkaline hydrolysis and low reactivity with amines [27]. Previously, porous structure in PHEMA hydrogels was obtained by phase separation using a low-molecular-weight or polymeric porogen, or by the salt-leaching method. The material was used as a mouse embryonic stem cell support [8, 28, 29]. The aim of this report is to demonstrate conditions under which communicating pores are formed enabling high permeability of PHEMA scaffolds, which is crucial for future cell seeding.

16.2 EXPERIMENTAL

16.2.1 REAGENTS

2-Hydroxyethyl methacrylate (HEMA; Röhm GmbH, Germany) and ethylene dimethacrylate (EDMA; Ugilor S.A., France), were purified by distillation. 2,2'-Azobisisobutyronitrile (AIBN, Fluka) was crystallized from ethanol and used as initiator. Sodium chloride G.R. (Lach-Ner, s.r.o. Neratovice, Czech Republic) was classified, particle size 250–500 μm and ammonium sulfate needles (100×600 μm, Lachema, Neratovice, Czech Republic) were used as porogens. Cyclohexanol (CyOH, Lachema, Neratovice, Czech Republic) was distilled, dodecan-1-ol (DOH) and all other solvents and reagents were obtained from Aldrich and used without purification. Ammonolyzed PGMA microspheres (2 μm) were obtained by the previously described procedure [30]. Sulfonated polystyrene (PSt) microspheres (8 μm) Ostion LG KS 0803 were purchased from Spolek pro chemickou a hutní výrobu, Ústí n. L., Czech Republic. Polyaniline hydrochloride microspheres (PANI, 200–400 nm) were prepared according to literature [31].

16.2.2 HYDROGEL PREPARATION

Cross-linked hydrogel constructs were prepared by the bulk radical polymerization of a reaction mixture containing monomer (HEMA), cross-linking agent (EDMA), initiator (AIBN) and NaCl or/and liquid diluent as a porogen (CyOH/DOH = 9/1 w/w). The compositions of polymerization mixtures are summarized in Table 16.1. The amount of cross linker (2 wt.%) and AIBN (1 wt.%) dissolved in monomers was the same in all experiments, while the amount of porogen in the polymerization batch was varied from 35.9–41.4 vol.%. Optionally, needle-like $(NH_4)_2SO_4$ crystals (42.3 vol.%) together with saturated $(NH_4)_2SO_4$ solution were used as a porogen instead of NaCl crystals, allowing thus formation of hydrogels with communicating pores (Run 9, Table 16.1). For the sake of comparison, a copolymer was prepared with a mixture of solid (NaCl) and liquid low-molecular-weight porogen (diluent), amounting to 50% of the polymerization feed (Run 8). The thickness of the hydrogel was adjusted with a 3-mm-thick silicone rubber spacer between the Teflon plates (10 10 cm), greased with a silicone oil and covered with Cellophane. The reaction mixture was transferred onto a hollow plate and covered with a second plate, clamped and heated at 70°C for 8 h. After polymerization, the hydrogels obtained with inorganic salt as a porogen were soaked in water and washed until the reaction of chloride or sulfate ions disappeared. The scaffolds prepared in the presence of a liquid diluent were washed with ethanol/water mixtures (98/2, 70/30, 40/60, 10/90 v/v) and water to remove the diluent, unreacted monomers and initiator residues. The washing water was then changed every day for 2 weeks.

TABLE 16.1 Preparation of PHEMA hydrogels, conditions and properties[a].

Run	NaCl (vol.%)	Water regain (ml/g)		CX regain[d] (ml/g)	Cumulative pore volume[f] (ml/g)
1	41.4	0.84[d]	7.52[e]	0.13	0.35
2	40.8	0.88[d]	7.40[e]	0.23	0.47
3	40.0	1.04[d]	5.34[e]	0.56	1.03
4	39.1	0.81[d]	4.05[e]	0.33	1.24
5	37.9	0.78[d]	4.04[e]	0.21	1.60
6	37.0	0.79[d]	3.64[e]	0.34	1.83
7	35.9	0.75[d]	3.32[e]	0.32	1.70
8[b]	37.9	0.89[d]	4.30[e]	0.45	1.65
9	42.3[c]	0.84[d]	2.11[e]	0.08	0.08

[a] Cross-linked with 2 wt.% EDMA, 1 wt.% AIBN, NaCl in vol.% relative to polymerization mixture (HEMA + EDMA + NaCl).

[b] One half of the HEMA/EDMA feed was replaced by CyOH/DOH = 9/1 w/w.

[c] $(NH_4)_2SO_4$.

[d] Centrifugation method.

[e] Suction method.

[f] Mercury porosimetry.

16.2.3   METHODS

16.2.3.1   MICROSCOPY

Low-vacuum scanning electron microscopy (LVSEM) was performed with a microscope Quanta 200 FEG (FEI, Czech Republic). Neat hydrated hydrogels were cut with a razor blade into ~ 5-mm cubes, flash-frozen in liquid nitrogen and placed on the sample stage cooled to −10°C. Before microscopic observation, the top of a frozen sample was cut off using a sharp blade. During the observation, the conditions in the microscope (–10°C, 100 Pa) caused slow sublimation of ice from the sample, which made it possible to visualize its 3D morphology. All samples were observed with a low-vacuum secondary electron detector, using the accelerating voltage 30 kV. Lyophilized PHEMA hydrogels Run 3 and Run 8 filled with microspheres were also examined by LVSEM; however, the microspheres were washed out of the pores during freezing and, consequently, they were scarcely observed on the micrographs.

High-vacuum scanning electron microscopy (HVSEM) was carried out with an electron microscope Vega TS 51355 (Tescan, Czech Republic). Permeability of the water-imbibed hydrogels (Runs 7 and 9) was investigated by the flow of water suspension of the polymer microspheres. Before observation, the wet hydrogel was placed on a wet filtration paper and a droplet of a suspension of 8-μm PSt microspheres in water was placed on the top. The sample was dried at ambient temperature and cut with a sharp blade in the direction of the microsphere flow. Samples showing the top, bottom and cross-sections of flowed-through hydrogels were sputtered with a 8-nm layer of platinum using a vacuum sputter coater (Baltec SCD 050), fixed with a conductive paste to a brass support and viewed in a scanning electron microscope in high vacuum (10^{-3} Pa), using the acceleration voltage 30 kV and a secondary-electrons detector. This technique made it possible to observe both microspheres and pores of the hydrogel.

16.2.3.2 SOLVENT REGAIN

The solvent (water or cyclohexane—CX) regain was determined in 1×2 cm sponge pieces of hydrogel kept for 1 week in deionized water, which was exchanged daily. Water regain was measured by two methods: (i) centrifugation [32] (WR_c) and (ii) suction (WR_s). In centrifugation method, solvent-swollen samples were placed into glass columns with fritted disc, centrifuged at 980 g for 10 min and immediately weighed (w_w – weight of hydrated sample), then vacuum-dried at 80°C for 7 h and again weighed (w_d – weight of dry sample). In the second method, excessive water was removed from the imbibed hydrogel by suction and the hydrogel weighted to determine w_w. Weight of dry sample w_d was determined as above. Water regains WR_c or WR_s (mL/g) were calculated according to the equation:

$$WR_c \left(WR_s \right) = \frac{w_w - w_d}{w_d} \qquad (1)$$

The results are average values of two measurements for each hydrogel. To measure cyclohexane regain (CXR) by centrifugation, equilibrium water-swollen hydrogels were successively washed with ethanol, acetone and finally cyclohexane. Using the solvent-exchange, a thermodynamically good (swelling) solvent in the swollen gel was replaced by a thermodynamically poor solvent (nonsolvent). Porosity of the hydrogels (p) was calculated from the water and cyclohexane regains (Table 16.1) and PHEMA density ($\rho = 1.3$ g/mL) according to the equation:

$$p = \frac{R \times 100}{R + \frac{1}{\rho}} \ (\%) \qquad (2)$$

where $R = WR_c$, WR_s, or CXR (mL/g).

16.2.3.3 MERCURY POROSIMETRY

Pore structure of freeze-dried PHEMA scaffolds was characterized on a mercury porosimeter Pascal 140 and 440 (Thermo Finigan, Rodano, Italy). It works in two pressure intervals, 0–400 kPa and 1–400 MPa, allowing determination of meso- (2–50 nm), macro (50–1000 nm) and small su-perpores (1–116 µm). The pore volume and most frequent pore diameter were calculated under the assumption of a cylindrical pore model by the PASCAL program. It employed Washburn's equation describing capillary flow in porous materials [33]. The volumes of bottle and spherical pores were evaluated as the difference between the end values on the volume/pressure curve. Porosity was calculated according to Eq. (2), where cumulative pore volume (meso-, macro and small superpores) from mercury porosimetry was used for R.

16.3 RESULTS AND DISCUSSION

16.3.1 MORPHOLOGY OF HYDROGELS

The prepared PHEMA constructs had always an opaque appearance indicating a permanent porous structure. Pores are generally divided into micro, meso-, macropores and small and large superpores. Morphology of water-swollen PHEMA hydrogels was investigated by LVSEM as shown in Fig. 16.1. Large 200–500 µm superpores were developed as imprints of NaCl crystals, which were subsequently washed out from the hydrogel; the interstitial space between them was filled with the polymer. During the observation, ice crystals filling soft polymer net were clearly visible in the center of the hydrogel Run 1 (Table 16.1) prepared at the highest content of NaCl (41.4 vol.%) in the feed (Fig. 16.1a). The internal surface area was too small to be determined. Fig. 16.1b and c show hydrogels from Run 3 and 5 (40 and 37.9 vol.% NaCl), respectively, documenting their more compact structure accompanied by thicker walls between large superpores as compared with the hydrogel from Run 1 (Fig. 16.1a). According to LVSEM, *ca.* 8-µm pores, the presence of which was confirmed by mercury porosimetry (volume about 1 mL/g), were observed in the walls between the large superpores. Longitudinal cracks in the material structure (Fig. 16.1b) were obviously caused by sample handling and fast freezing in liquid nitrogen. Nevertheless, the LVSEM micrographs displayed only cross-sections of hydrogels and it was not clear whether their pores are interconnected.

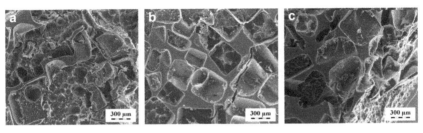

FIGURE 16.1   LVSEM micrographs showing frozen cross-section of PHEMA hydrogels prepared with (a) 41.4 vol.%, Run 1, (b) 40 vol.%, Run 3, and (c) 37.9 vol.% NaCl (250–500 μm), Run 5. PHEMA cross-linked with 2 wt.% EDMA (relative to monomers).

Interconnection of pores is of vital importance for cell ingrowths in future applications to tissue regeneration. This feature was tested by the permeability of the whole hydrogels for different kinds of microspheres under two microscopic observations. First, cross-sections of the frozen hydrogels filled with microspheres were observed in LVSEM. Second, the water-swollen hydrogels were flowed through by a suspension of microspheres in water and their dried cross-sections were then viewed in HVSEM.

LVSEM showed water-swollen morphology of hydrogel constructs, which preserved due to their freezing in liquid nitrogen. PHEMA Run 3 prepared with neat NaCl (Fig. 16.2a) was compared with Run 8 obtained in the presence of NaCl together with a mixture of CyOH/DOH (Fig. 16.2d). Addition of liquid porogens did not change the morphology; however, it increased the pore volume (from 0.21 to 0.45 mL CX/g, Table 16.1) and softness of the hydrogel. As a result, it had a tendency to disintegrate during the washing procedure. LVSEM of both PHEMA hydrogels filled with 2-μm ammonolyzed PGMA and 200–400 nm PANI microspheres is illustrated in Fig. 16.2b, e and Fig. 16.2c, f, respectively. The micrographs showed undistorted morphology of the frozen hydrogels, but just a few microspheres and/or their agglomerates. This was attributed to the fact that most of them were washed out during preparation of the sample for LVSEM.

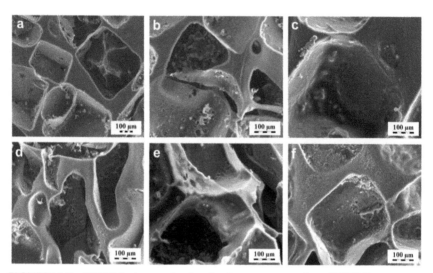

FIGURE 16.2   LVSEM micrographs showing frozen cross-section of PHEMA constructs; (a–c) Run 3, (d–f) Run 8; (a, d) neat and filled with (b, e) 2-μm ammonolyzed PGMA and (c, f) 200–400 nm PANI microspheres.

Morphology of the PHEMA hydrogels flowed through by a suspension of microspheres was observed by HVSEM. Figure 16.3 shows HVSEM micrographs of cross-sections of the top and bottom part of the hydrogels from Runs 3, 5 and 8 flowed through by a suspension of 8-μm sulfonated PSt microspheres in water. While the microspheres flowed through the hydrogel construct from Run 3, they did not penetrate the ones from Run 5 and 8 prepared in the presence of a rather low content of NaCl (37.9 vol.%). At the same time, surface and inner structure of the hydrogels slightly differed. Figure 16.4 shows HVSEM of the longitudinal section of the PHEMA scaffold obtained with cubic NaCl crystals as a porogen (Run 7). While Fig. 16.4a shows the bulk, Fig. 16.4b–d detailed sections. Again, small superpores with an average size about 13 μm were in the walls between the large superpores, forming small channels through which water flowed. To prove or exclude the interconnection of at least some pores, a suspension of 8-μm sulfonated polystyrene (PSt) microspheres in water was poured on the center of the topside of the gel. While water flowed through the hydrogel bulk, the microspheres were retained on the surface of the hydrogel or penetrated only superficial layers due to the surface cracks (Fig. 16.4 a,b). This confirmed that the pores of PHEMA hydrogels obtained with a low content of NaCl porogen (35.9 vol.%) did not communicate. In contrast, Fig. 16.5 presents longitudinal section of the PHEMA hydrogel from Run 9 (both bulk and detailed) obtained with needle-like $(NH_4)_2SO_4$ crystals as a porogen. This porogen allowed formation of connected pores, which is explained by the needle-like structure of ammonium sulfate crystals that are linked to the gel structure. At the same time, the crystals grew to large structures due to the presence of saturated $(NH_4)_2SO_4$ solution in the feed. As a result, long interconnected large superpores—channels—were formed. This is documented in Fig. 16.5a–d by the fact that suspension of 8-μm sulfonated PSt microspheres in water deposited in the center of the topside of the hydrogel flowed through. The captured microspheres are well visible in Fig. 16.5b–d. They accumulated at the places of pore narrowing; their majority, however, was found on the bottom part of the hydrogel. In such a way, the flow of water suspension of microspheres in the hydrogel was traced.

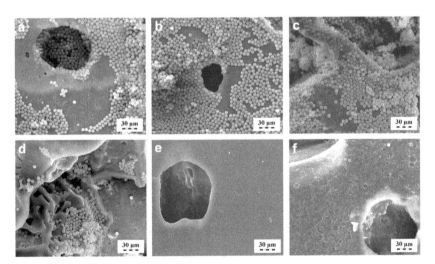

FIGURE 16.3 HVSEM micrographs of PHEMA hydrogels Run 3 (a, d), Run 5 (b, e) and Run 8 (c, f) showing top (a–c) and bottom (d–f) of the hydrogels after the flow of a suspension of 8-μm sulfonated PSt microspheres in water.

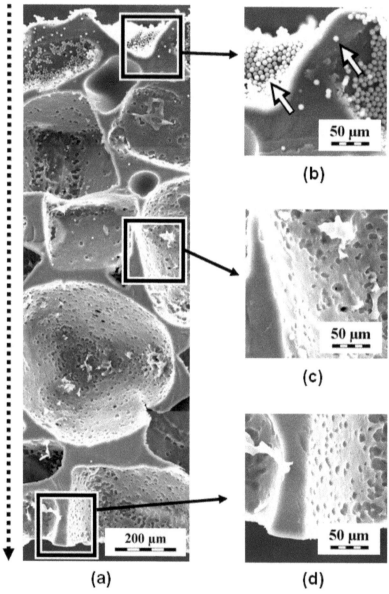

FIGURE 16.4   Selected HVSEM micrographs showing longitudinal section of PHEMA hydrogel 3 mm thick (Run 7) obtained with NaCl (250–500 μm) as a porogen after passing of a suspension of 8-μm sulfonated PSt microspheres in water (in the direction of the dotted line). (a) The whole cross-section through the hydrogel and selected details from (b) top, (c) center and (d) bottom. PSt microspheres are denoted with white arrows.

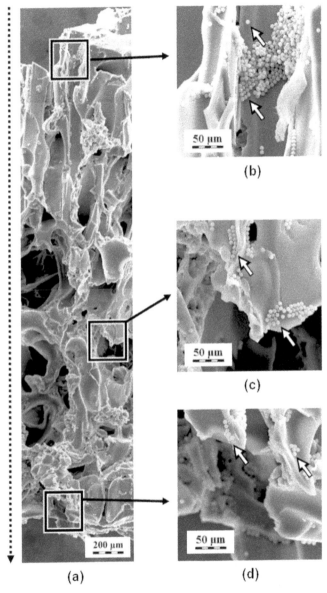

FIGURE 16.5 Selected HVSEM micrographs showing longitudinal section of PHEMA hydrogel 3 mm thick (Run 9) obtained with $(NH_4)_2SO_4$ (100×600 µm) as a porogen after passing of a suspension of 8-µm sulfonated PSt microspheres in water (in the direction of the dotted line). (a) The whole section through the construct and selected details from (b) top, (c) center and (d) bottom. PSt microspheres are denoted with white arrows.

Mechanical properties of the porous constructs were sensitive to the concentration of porogen in the feed. Hydrogels with lower contents of NaCl and therefore higher proportion of PHEMA had thicker walls between the pores and were more compact allowing increased swelling of polymer chains in water. Two PHEMA hydrogels with the highest contents of NaCl in the feed (41.4 vol.%—Run 1 and 40.8 vol.%—Run 2) possessing thin polymer walls between large superpores easily disintegrated as well as hydrogel prepared using $(NH_4)_2SO_4$ (42.3 vol.%—Run 9).

16.3.2 CHARACTERIZATION OF POROSITY BY SOLVENT REGAIN

Dependences of porosity of PHEMA hydrogels calculated from water or cyclohexane regain and also from mercury porosimetry on NaCl content in the polymerization feed showed similar behavior (Fig. 16.6). Porosities 81–91 and 49–57% for water regain were obtained by suction and centrifugation, respectively, 14–42% for cyclohexane regain and 31–70% for mercury porosimetry. The porosity determined by centrifugation of samples soaked with water and cyclohexane (solvents with different affinities to polar methacrylate chain) consists of two contributions: filling of the pores and swelling (solvation) of PHEMA chains. The uptake of cyclohexane, a thermodynamically poor solvent that cannot swell the polymer, is a result of the former contribution only, reflecting thus the pore volume. The water regain from centrifugation was always higher than the cyclohexane regain demonstrating thus swelling of polymer chains with water (Table 16.1). Solvent regains were affected by the concentration of NaCl porogen in the polymerization feed. Porosities according to both water and cyclohexane regains by centrifugation slightly increased with increasing volume of NaCl porogen in the polymerization feed from 35.9 to 40 vol.% and then decreased with a further NaCl increase up to 41.4 vol.% (Fig. 16.6). In the latter range of NaCl, the porosity evaluated by mercury porosimetry exhibited an analogous dependence. This decrease in solvent and mercury regains can be explained by thin polymer walls between large superpores inducing collapse of the porous structure. In the concentration range of NaCl in the feed 35.9–40 vol.%, mercury porosimetry provided higher porosities than those obtained from regains by centrifugation at 980 g because it obviously did not retain solvents in large superpores. Retained water reflected thus only small superpores, closed pores and solvation of the polymer in water similarly as observed earlier for macroporous PHEMA scaffolds [34]. Water regain was determined also by the suction method (Table 16.1) which gave the values several times higher (3.3–7.5 mL/g) than by centrifugation due to filling all the pores in the polymer structure, including large superpores. As expected, porosity by the suction method increased with increasing volume of NaCl porogen in the polymerization feed (Fig. 16.6).

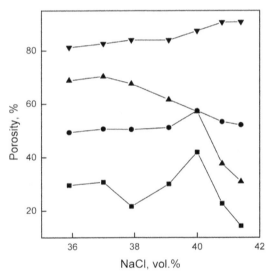

FIGURE 16.6 Dependence of porosity of PHEMA hydrogels determined from cyclohexane (■) and water regain measured by centrifugation (●) or suction (▼) and mercury porosimetry (▲) on the content of NaCl (250–500 μm) porogen in the polymerization feed.

The hydrogel from Run 8 formed in the presence of NaCl and CyOH/DOH porogen showed higher solvent regains and mercury penetration than the comparable hydrogel from Run 5 obtained with the same content of neat NaCl (Table 16.1). This can be explained by the higher total amount of porogen in the former hydrogel. In contrast, the hydrogel from

Run 9 prepared with needle-like $(NH_4)_2SO_4$ crystals as a porogen had the lowest solvent and mercury regains of all the samples. The exception was water regain by centrifugation, which was identical with that of sample Run 1 (Table 16.1) having a similar content of the NaCl porogen in the feed. This can imply that only large continuous superpores were present in this hydrogel and small superpores, macro and mesopores were almost absent as evidenced by the low values of solvent and mercury regains.

16.3.3   CHARACTERIZATION OF POROSITY BY MERCURY POROSIMETRY

The advantage of mercury porosimetry is that it provides not only pore volumes, but also pore size distribution not available by other techniques. The method measures samples dried by lyophilization, which does not distort the pore structure. As already mentioned, porosities determined by mercury porosimetry were lower than those obtained from water regain by the suction, which included large superpores, and higher than those from water and cyclohexane regain detected by centrifugation. This was due to better filling of the compact xerogel structures obtained at lower contents of NaCl in the feed with mercury under a high pressure than with water or cyclohexane under atmospheric pressure. Figure 16.7 shows the dependence of most frequent mesopore size of PHEMA scaffolds and their pore volumes on the NaCl porogen content in the polymerization feed. Predominantly, 4–5 nm mesopores were detected with their volume increasing from 0.03 to 0.1 mL/g with increasing NaCl content in the polymerization feed. Macropores were absent and very low values of specific surface areas (<0.1 m^2/g) were found. The presence of CyOH/DOH porogen (Run 8) did not substantially affect the formation of meso- and macropores (volume 0.022 mL/g), because the amount of cross linker in the polymerization feed was limited to only 2 wt.%. The separation of the polymer from the porogen phase could not thus occur and porous structure was not formed. Both in hydrophobic styrene-divinylbenzene [35, 36] and polar methacrylate copolymers [37] prepared in the presence of liquid porogens, phase separation and formation of macroporous structure occurred at cross linker contents higher than 10 wt.%.

Figure 16.8 represents the dependence of most frequent small superpore size of PHEMA constructs and their pore volume on the content of NaCl porogen in the polymerization. Pore size increased up to 28–69 μm with increasing NaCl volume. This was pronounced in the range 40–41.4 vol.% NaCl probably due to the aggregation of NaCl crystals in the mixture at their high contents. All the investigated samples contained small superpores, the volume of which was about 20 times higher than that of mesopores. The volume of small superpores continuously decreased from 1.8 to 0.2 mL/g with raising NaCl amount in the feed. This could be explained by collapse of the pore structure and destruction of last two hydrogels with the highest content of NaCl porogen (Runs 1 and 2) under high pressure as mentioned above. The size of small superpores according to mercury porosimetry was by an order of magnitude smaller than the particle size of the used NaCl porogen (250–500 μm) because the method was able to distinguish the superpores only in the size range 1–116 μm (Fig. 16.9). Large superpores (imprints of NaCl crystals) were detected by LVSEM. Figure 16.9 exemplifies a typical cumulative pore volume and a derivative pore size distribution curve of PHEMA Run 4 with a decisive contribution of small superpores 25 μm in size.

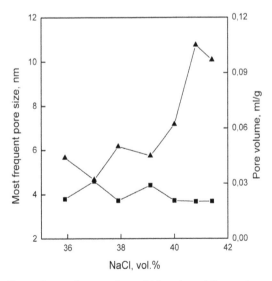

FIGURE 16.7   Dependence of pore volume (▲) and most frequent mesopore size (■) of PHEMA hydrogels on the content of NaCl (250–500 μm) porogen in the polymerization feed according to mercury porosimetry.

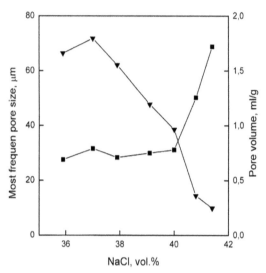

FIGURE 16.8 Dependence of pore volume (▼) and most frequent small superpore size (■) of PHEMA hydrogels on the content of NaCl (250–500 µm) porogen in the polymerization feed according to mercury porosimetry.

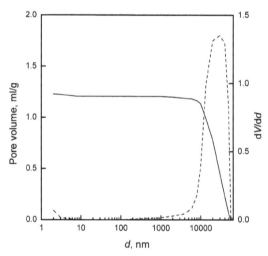

FIG.URE16.9 Cumulative pore volume (–) and pore size distribution (–) of the Run 4 hydrogel determined by mercury porosimetry in the range 1.88 nm–116 µm.

16.4 CONCLUSIONS

Superporous PHEMA constructs were prepared by bulk radical copolymerization of HEMA and EDMA in the presence of NaCl or/and liquid diluent (CyOH/DOH) or $(NH_4)_2SO_4$ crystals. Morphology of the prepared scaffolds was characterized by several methods including scanning electron microscopy both in swollen (LVSEM) and dry (HVSEM) state, solvent (water and cyclohexane) regains, high- and low-pressure mercury porosimetry of lyophilized samples and dynamic desorption of nitrogen. Morphology and porous structure of the hydrogels were preferentially affected by the character and amount of the used porogen—NaCl, CyOH/DOH mixture or $(NH_4)_2SO_4$. After washing out of the salts and solvents from PHEMA, three types of pores were detected by microscopic and mercury porosimetry methods, including large superpores (hundreds of micrometers) as imprints of salt crystals. The hydrogels formed can be divided into two groups, with disconnected and interconnected pores. The latter allowed the passage of suspension of microspheres in water, which was observed only for the samples with ammonium sulfate and the highest content of NaCl crystals used as a porogen in the feed. Interconnected pores are crucial for potential application of the scaffolds as living cell supports. LVSEM showed the undistorted (frozen) structure of the hydrogels, but only few flowed-through microspheres could be observed as they tended to escape from the pores during sample preparation. HVSEM seemed to be the best microscopic technique especially for viewing permeability of hydrogels to 8-μm microspheres. Hydrogels were initially flowed through by the particles in their natural wet state, but the specimens were then dried before SEM observation. The microparticles could be traced both on the upper/lower parts of the hydrogels and on the longitudinal sections.

Mercury porosimetry provided detailed description of morphology of PHEMA constructs with pore sizes from units of nanometers to tens of micrometers. The drawback of the method is that the hydrogels are not measured in the swollen, but dry state, as xerogels. But comparison of the data in both wet and dry states showed that lyophilization did not change the pore structure. The mesopores and small superpores detected by mercury porosimetry cannot be formed by the imprinting mechanism. While mesopores present only in very small amounts may be formed by phase separation, small superpores arise by polymer contraction in the walls of large superpores. Small 28–69 μm superpores were mainly present in the porous structure apart from the large superpores (200–500 μm imprints of solid porogen crystals), the volume of which was several times higher than that of other pores, as confirmed by water regain obtained by the suction method.

ACKNOWLEDGMENT

Financial support of the Grant Agency of the Czech Republic (project No. P304/11/0731) is gratefully acknowledged.

KEYWORDS

- design and synthesis
- ethylene dimethacrylate (EDMA)
- hydrogel-based supports
- poly(2-hydroxyethyl methacrylate) (PHEMA)
- polymerization of 2-hydroxyethyl methacrylate (HEMA)
- research methodology

REFERENCES

1. Kofron, M. D., Cooper, J. A., Kumbar, S. G., Laurencin, C. T., Novel tubular composite matrix for bone repair. J Biomed Mater Res, Part A 2007; 82: 415–425.
2. Moroni, L., Hendriks, J. A. A., Schotel, R., De Wijn, J. R., Van Blitterswijk, C. A., Design of biphasic polymeric 3-dimensional fiber deposited scaffolds for cartilage tissue engineering applications. Tissue Eng 2007; 13: 361–371.
3. Dvořánková, B., Holíková, Z., Vacík, J., Königová, R., Kapounková, Z., Michálek, J., Přádný, M., Smetana, K., Reconstruction of epidermis by grafting of keratinocytes cultured on polymer support clinical study. Int J Dermatol 2003; 42: 219–223.
4. Bhang SH, Lim JS, Choi CY, Kwon YK, Kim BS. The behavior of neural stem cells on biodegradable synthetic polymers. J Biomater Sci, Polym Ed 2007; 18: 223–239.

5. Bianchi F, Vassalle C, Simonetti M, Vozzi G, Domenici C, Ahluwalia A. Endothelial cell function on 2D and 3D microfabricated polymer scaffolds: Applications in cardiovascular tissue engineering. J Biomater Sci, Polym Ed 2006; 17: 37–51.

6. Sander EA, Alb AM, Nauman EA, Reed WF, Dee, K. C., Solvent effects on the microstructure and properties of 75/25 poly(D, L-lactide-coglycolide) tissue scaffolds. J Biomed Mater Res, Part A 2004; 70: 506–513.

7. Nam, Y. S., Yoon, J. J., Park, T. G., A novel fabrication method of macroporous biodegradable polymer scaffolds using gas foaming salt as a porogen additive. J Biomed Mater Res, Appl Biomater 2000; 53: 1–7.

8. Kroupová, J., Horák, D., Pacherník, J., Dvořák, P., Šlouf, M., Functional polymer hydrogels for embryonic stem cell support. J Biomed Mater Res, Part B: Appl Biomater 2006; 76B: 315–325.

9. Tighe, B., Corkhill, P., Hydrogels in biomaterials design: Is there life after polyHEMA? Macromol Rep 1994; A31, 707–713.

10. Plieva, F. M., Galaev, I. Y., Mattiasson, B., Macroporous gels prepared at subzero temperatures as novel materials for chromatography of particulate-containing fluids and cell culture applications. J Sep Sci 2007; 30: 1657–1671.

11. Mooney, D. J., Baldwin, D. F., Suh, N. P., Vacanti, J. P., Langer, R., Novel approach to fabricate porous sponges of poly(D, L-lactic-coglycolic acid) without the use of organic solvents. Biomaterials 1996; 17, 1417–1422.

12. Chung, H. J., Park, T. G., Surface engineered and drug releasing prefabricated scaffolds for tissue engineering. Adv Drug Delivery Rev 2007; 59: 249–262.

13. Ferreira, L. S., Gerecht, S., Fuller, J., Shieh, H. F., Vunjak-Novakovic, G., Langer, R., Bioactive hydrogel scaffolds for controllable vascular differentiation of human embryonic stem cells. Biomaterials 2007; 28: 2706–2717.

14. Tangsadthakun, C., Kanokpanont, S., Sanchavanakit, N., Pichyangkura, R., Banaprasert, T., Tabata, Y., Damrongsakkul, S., The influence of molecular weight of chitosan on the physical and biological properties of collagen/chitosan scaffolds. J Biomater Sci, Polym Ed 2007; 18: 147–163.

15. Wachiralarpphaithoon, C., Iwasaki, Y., Akiyoshi, K., Enzyme-degradable phosphorylcholine porous hydrogels cross-linked with polyphosphoesters for cell matrice. Biomaterials 2007; 28: 984–993.

16. **Treml, H., Woelki, S., Kohler, H. H.,** Theory of capillary formation in alginate gels. Chem Phys 2003; 293: 341–353.

17. Konno, T., Ishihara, K., Temporal and spatially controllable cell encapsulation using a water-soluble phospholipid polymer with phenylboronic acid moiety. Biomaterials 2007; 28: 1770–1777.

18. Heckmann, L., Schlenker, H. J., Fiedler, J., Brenner, R., Dauner, M., Bergenthal, G., Mattes, T., Claes, L., Ignatius, A., Human mesenchymal progenitor cell responses to a novel textured poly(L-lactide) scaffold for ligament tissue engineering. J Biomed Mater Res, Part, B., Appl Biomater 2007; 81: 82–90.

19. Darling AL, Sun W. 3D microtomographic characterization of precision extruded poly ε-caprolactone scaffolds. J Biomed Mater Res, Part, B., Appl Biomater 2004; 70: 311–317.

20. Rhee, W., Rosenblatt, J., Castro, M., Schroeder, J., Rao, P. R., Harner, C. F. H., Berg, R. A., In vivo stability of poly(ethylene glycol)-collagen composites, in: Poly(Ethylene

Glycol) Chemistry and Biological Applications, Harris, J. M., Zalipsky, S., Eds, ACS Symp Ser 1997; 680: 420–440.

21. Savina, I. N., Galaev, I. Y., Mattiasson, B., Ion-exchange macroporous hydrophilic gel monolith with grafted polymer brushes. J Mol Recognit 2006; 19: 313–321.

22. Carampin, P., Conconi, M. T., Lora, S., Menti, A. M., Baiguera, S., Bellini, S., Grandi, C., Parnigotto Electrospun polyphosphazene nanofibers for in vitro rat endothelial cells proliferation. J Biomed Mater Res, Part A 2007; 80: 661–668.

23. Zhang, C. H., Zhang, N., Wen, X. J., Synthesis and characterization of biocompatible, degradable, light-curable, polyurethane-based elastic hydrogels. J Biomed Mater Res, Part A 2007; 82: 637–650.

24. Castner, D. G., Ratner, B. D., Biomedical surface science: Foundation to frontiers. Surf Sci 2002; 500: 28–60.

25. Lee, K. Y., Mooney, D. J., Hydrogels for tissue engineering. Chem Rev 2001; 101: 1869–1879.

26. Refojo, M. F., Hydrophobic interactions in poly(2-hydroxyethyl methacrylate) homogeneous hydrogel. J Polym Sci, Part A1, Polym Chem 1967; 5: 3103–8.

27. Ratner, B. D., Hoffman, A. S., Hydrogels for Medical and Related Applications. ACS Symp Ser 1976; 31: 1–36.

28. Horák, D., Dvořák, P., Hampl, A., Šlouf, M., Poly(2-hydroxyethyl methacrylate-*co*-ethylene dimethacrylate) as a mouse embryonic stem cell support, J Appl Polym Sci 2003; 87, 425–432.

29. Horák, D., Kroupová, J., Šlouf, M., Dvořák, P., Poly(2-hydroxyethyl methacrylate)-based slabs as a mouse embryonic stem cell support. Biomaterials 2004; 25: 5249–5260.

30. Horák, D., Shapoval, P., Reactive poly(glycidyl methacrylate) microspheres prepared by dispersion polymerization. J Polym Sci, Part, A., Polym Chem Ed 2000; 38: 3855–3863.

31. Stejskal, J., Kratochvíl, P., Gospodinova, N., Terlemezyan, L., Mokreva, P., Polyaniline dispersions: Preparation of spherical particles and their light-scattering characterization. Polymer 1992; 33: 4857–4858.

32. Štamberg, J., Ševčík, S., Chemical transformations of polymers, III, Selective hydrolysis of a copolymer of diethylene glycol methacrylate and diethylene glycol dimethacrylate. Collect Czech Chem Commun 1966; 31: 1009–2016.

33. Porosimeter Pascal 140 and Pascal 440, Instruction manual, p. 8.

34. Hradil, J., Horák, D., Characterization of pore structure of PHEMA-based slabs. React Funct Polym 2005; 62: 1–9.

35. Millar, J. A., Smith, D. G., Marr, W. E., Kresmann, T. R. E., Solvent modified polymer networks. Part 1. The preparation and characterization of expanded-networks and macroporous styrene-DVB copolymers and their sulfonates. J Chem Soc 1963; 218–225.

36. Kun, K. A., Kunin, R., Macroreticular resins, III, Formation of macroreticular styrene-divinylbenzene copolymers. J Polym Sci, Part A1, Polym Chem 1968; 6: 2689–2701.

37. Hradil, J., Křiváková, M., Starý, P., Čoupek, J., Chromatographic properties of macroporous copolymers of 2-hydroxyethyl methacrylate and ethylene dimethacrylate. J Chromatogr 1973; 79: 99–105.

INDEX